Raspberry Pi Robotic Blueprints

Utilize the powerful ingredients of Raspberry Pi to bring to life amazing robots that can act, draw, and have fun with laser tag

Richard Grimmett

BIRMINGHAM - MUMBAI

Raspberry Pi Robotic Blueprints

First published: October 2015

Production reference: 1261015

Published by Packt Publishing Ltd.
Livery Place
35 Livery Street
Birmingham B3 2PB, UK.

ISBN 978-1-78439-628-2

www.packtpub.com

Credits

Author
Richard Grimmett

Reviewers
James McNutt
Werner Ziegelwanger, MSc

Commissioning Editor
Sarah Crofton

Acquisition Editor
Tushar Gupta

Content Development Editor
Kirti Patil

Technical Editor
Manthan Raja

Copy Editor
Vibha Shukla

Project Coordinator
Kranti Berde

Proofreader
Safis Editing

Indexer
Monica Ajmera Mehta

Graphics
Disha Haria

Production Coordinator
Conidon Miranda

Cover Work
Conidon Miranda

About the Author

Richard Grimmett continues to have more fun than he should be allowed working on robotics projects while teaching computer science and electrical engineering at Brigham Young University-Idaho. He has a bachelor's and master's degree in electrical engineering and a PhD in leadership studies. This is the latest book, in a long series of books detailing how to use Raspberry Pi, Arduino, and BeagleBone Black for robotics projects, written by him.

About the Reviewers

James McNutt first got his hands on Raspberry Pi while writing the curriculum for a library summer program that was designed to teach teens the basics of web design, robotics, and programming. He has incorporated Raspberry Pis into several of his projects; however, none have been as meaningful as those focused on education. There is nothing as effective at breaking down students' trepidation around computer science as placing Raspberry Pi in their hands and assuring them that they won't break it and—even if they do break it—that they're going to learn something in the process. Inspired by the unique way in which libraries touch the lives of their patrons and communities, James has continued his involvement with libraries, working and teaching at some of the country's first library makerspaces. Now, working as a library systems administrator, James still sets aside the time to teach public classes at his library.

I'd like to acknowledge Meg Backus, Bill Brock, Lindsey Frost, and Justin Hoenke for their commitment to providing educational opportunities in libraries and for what they have personally taught me.

Werner Ziegelwanger, MSc has studied game engineering and simulation and obtained his master's degree in 2011. His master's thesis, *Terrain Rendering with Geometry Clipmaps for Games, Diplomica Verlag*, was published. His hobbies include programming, gaming, and exploring all kinds of technical gadgets.

Werner worked as a self-employed programmer for many years and mainly did web projects. During this time, he started his own blog (https://developer-blog.net/), which is about the Raspberry Pi, Linux, and open source.

Since 2013, Werner has been working as a Magento developer and is the head of programming at mStage GmbH, an e-commerce company that focuses on Magento.

www.PacktPub.com

Support files, eBooks, discount offers, and more

For support files and downloads related to your book, please visit www.PacktPub.com.

Did you know that Packt offers eBook versions of every book published, with PDF and ePub files available? You can upgrade to the eBook version at www.PacktPub.com and as a print book customer, you are entitled to a discount on the eBook copy. Get in touch with us at service@packtpub.com for more details.

At www.PacktPub.com, you can also read a collection of free technical articles, sign up for a range of free newsletters and receive exclusive discounts and offers on Packt books and eBooks.

https://www2.packtpub.com/books/subscription/packtlib

Do you need instant solutions to your IT questions? PacktLib is Packt's online digital book library. Here, you can search, access, and read Packt's entire library of books.

Why subscribe?

- Fully searchable across every book published by Packt
- Copy and paste, print, and bookmark content
- On demand and accessible via a web browser

Free access for Packt account holders

If you have an account with Packt at www.PacktPub.com, you can use this to access PacktLib today and view 9 entirely free books. Simply use your login credentials for immediate access.

Table of Contents

Preface

Robotics have really come into the public spotlight in the past few years. Ideas that, just a few years ago, would have lived only in the government research center or university lab, such as robotic vacuum cleaners, drones that cover the sky, and self-driving cars, are now making their way into everyday life. This movement is fueled, at least in part, by scores of enterprising individuals, without significant technical training, who undertake building their idea with inexpensive hardware and free, open-source software.

This book celebrates this effort by detailing how to get started on building the project that you always wanted to build but didn't think you had the expertise for. The heart of these projects is Raspberry Pi B version 2, a cable microprocessor-based system that can run Linux and provides a platform for a significant number of open source modules. Combine Raspberry Pi with these open source modules and low cost hardware, and you can build robots that can walk, role, draw, and even fly.

What this book covers

Chapter 1, Adding Raspberry Pi to an RC Vehicle, shows you how to add Raspberry Pi to an existing toy, such as an old RC car or truck, to make it "new" again.

Chapter 2, Adding Raspberry Pi to a Humanoid Robot, covers how to add Raspberry Pi to robots, such as the Robosapien line from WowWee, to add voice commands and make them more versatile.

Chapter 3, Building a Tracked Vehicle That Can Plan Its Own Path, explains how to build a tracked robot containing sensors so that it can map the position of a set of objects.

Chapter 4, Building a Robot That Can Play Laser Tag, covers how to use the capabilities of Raspberry Pi to build a wheeled robot that can play laser tag.

Chapter 5, A Robot That Can Draw, introduces the capability of external dedicated servo controllers that can make controlling the arms and legs of the robot much easier. This is done using servos, whose position can be controlled using our system.

Chapter 6, A Robot That Can Play Air Hockey, explains how to use stepper motors and an advanced vision system to build a robot that can plan air hockey using more power and precision.

Chapter 7, A Robot That Can Fly, explains that after building a robot that can walk, talk, or play air hockey, you can build a robot that can fly.

What you need for this book

Chapter	Software	Where Located
Chapter 1	Raspberry Pi Debian	`https://www.raspberrypi.org/`
	RaspiRobot Board V2 drivers from Simon Monk	`http://www.monkmakes.com/?page_id=698`
	TightVNC Server	`sudo apt-get install tightvncserver`
	luvcview	`sudo apt-get install luvcview`
Chapter 2	Arduino IDE	`https://www.arduino.cc/`
	eSpeak	`sudo apt-get install espeak`
	PocketSphinx	`http://cmusphinx.sourceforge.net/`
Chapter 3	RaspiRobot Board V2 drivers from Simon Monk	`http://www.monkmakes.com/?page_id=698`
Chapter 4	PodSixNet	`http://mccormick.cx/projects/PodSixNet/`
Chapter 5	Pololu Maestro Control Center	`http://www.pololu.com/docs/0J40/3.a`
Chapter 6	Arduino IDE	`https://www.arduino.cc/`
	OpenCV	`http://opencv.org/`

Who this book is for

This all-embracing guide is created for anyone who is interested in expanding their horizon in applying the peripherals of Raspberry Pi. If you fancy building complex-looking robots with simple, inexpensive, and readily available hardware, then this is the ideal book for you. Prior understanding of Raspberry Pi with simple mechanical systems is recommended.

Conventions

In this book, you will find a number of text styles that distinguish between different kinds of information. Here are some examples of these styles and an explanation of their meaning.

Code words in text, database table names, folder names, filenames, file extensions, pathnames, dummy URLs, user input, and Twitter handles are shown as follows: "Type cd rrb2-1.1 — this will change the directory to the location of the files."

A block of code is set as follows:

```
void loop()
{
   int dt;
   uint8_t logOutput=0;
   debug_counter++;
   timer_value = micros();
```

Any command-line input or output is written as follows:

```
volatile int viRobsapienCmd = -1;  // A robosapien command
  sent over the UART request
// Some but not all RS commands are defined
#define RSTurnRight        0x80
#define RSRightArmUp       0x81
```

New terms and **important words** are shown in bold. Words that you see on the screen, for example, in menus or dialog boxes, appear in the text like this: "Now click on **Connect** on **Remote Desktop Viewer**."

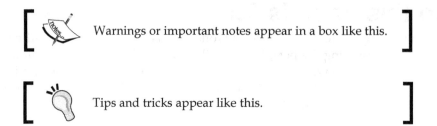

Warnings or important notes appear in a box like this.

Tips and tricks appear like this.

Reader feedback

Feedback from our readers is always welcome. Let us know what you think about this book—what you liked or disliked. Reader feedback is important for us as it helps us develop titles that you will really get the most out of.

To send us general feedback, simply e-mail feedback@packtpub.com, and mention the book's title in the subject of your message.

If there is a topic that you have expertise in and you are interested in either writing or contributing to a book, see our author guide at www.packtpub.com/authors.

Customer support

Now that you are the proud owner of a Packt book, we have a number of things to help you to get the most from your purchase.

Downloading the example code

You can download the example code files from your account at http://www.packtpub.com for all the Packt Publishing books you have purchased. If you purchased this book elsewhere, you can visit http://www.packtpub.com/support and register to have the files e-mailed directly to you.

Downloading the color images of this book

We also provide you with a PDF file that has color images of the screenshots/diagrams used in this book. The color images will help you better understand the changes in the output. You can download this file from https://www.packtpub.com/sites/default/files/downloads/6282OT_ColorImages.pdf.

Errata

Although we have taken every care to ensure the accuracy of our content, mistakes do happen. If you find a mistake in one of our books—maybe a mistake in the text or the code—we would be grateful if you could report this to us. By doing so, you can save other readers from frustration and help us improve subsequent versions of this book. If you find any errata, please report them by visiting http://www.packtpub.com/submit-errata, selecting your book, clicking on the **Errata Submission Form** link, and entering the details of your errata. Once your errata are verified, your submission will be accepted and the errata will be uploaded to our website or added to any list of existing errata under the Errata section of that title.

To view the previously submitted errata, go to https://www.packtpub.com/books/content/support and enter the name of the book in the search field. The required information will appear under the **Errata** section.

Piracy

Piracy of copyrighted material on the Internet is an ongoing problem across all media. At Packt, we take the protection of our copyright and licenses very seriously. If you come across any illegal copies of our works in any form on the Internet, please provide us with the location address or website name immediately so that we can pursue a remedy.

Please contact us at copyright@packtpub.com with a link to the suspected pirated material.

We appreciate your help in protecting our authors and our ability to bring you valuable content.

Questions

If you have a problem with any aspect of this book, you can contact us at questions@packtpub.com, and we will do our best to address the problem.

1
Adding Raspberry Pi to an RC Vehicle

The introduction of powerful, inexpensive processors that also provide a wide range of functionality through free open-source software has caused the do-it-yourself electronic project work to expand far beyond the simple, less than inspiring projects of the past. Now the developers can, with very low cost, create amazingly complex projects that were almost unthinkable a few years ago.

Many in this community are using Raspberry Pi as the basis for this revolution. This book provides simple, easy-to-follow instructions on how to use the Raspberry Pi in some very complex and sophisticated projects. Now enough of the introduction, let's start building something.

In this chapter, you'll learn the following:

- How to modify an Xmods RC car using Raspberry Pi
- How to set break into the control circuitry of the car and use Raspberry Pi to control it
- How to use wireless communication to add remote control to the car

Configuring Raspberry Pi – The brain of your projects

A brief note before you start. In this book, you'll be using Raspberry Pi B2, a microprocessor that can run on the Linux operating system. The following is an image of the unit, with the different interconnectors labeled:

As this is an advanced projects book, you have already spent some time with Raspberry Pi and know how to write Raspbian/Wheezy on an SD card and boot your Raspberry Pi. If you don't, feel free to go to the Raspberry Pi website at https://www.raspberrypi.org/. Here you'll find all the instructions that you need to get your Raspberry Pi B 2 up and running.

Note that you may want to install your system on a microSD card that has at least 8 GB of memory. In some of the projects that you'll be building, you'll be installing some fairly significant pieces of open source software and you may not want to run out of memory.

Now you are ready to start with some simple product modification. Let's start with an RC car; you'll replace the transmitter and control the car with a wireless connection on Raspberry Pi.

Configuring and controlling an RC car with Raspberry Pi

The first project that you'll be working on is a simple RC car, like the one shown here:

This particular car is an Xmods car, sold by Radio Shack, also available at other retail and online outlets. You can certainly use other RC cars as well. The advantage of this particular set is that the inputs to the drivetrain and steering are very easy to access.

The following is the car, exposing the center control mechanism:

There are two connections that you will want direct access to. The first is the drive motor, and the second is the steering mechanism. For this particular model of RC car, the drive mechanism is in the rear. What you are normally looking for is two wires that will directly drive the DC motor of the car. On this system, there is a connector in the rear of the car, it looks as shown in the following image:

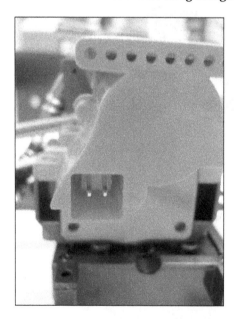

In the main control section of the car, you can see that there is a connector that plugs in these two wires in order to control the speed of the car, as shown here:

Remove this plug and these wires; you'll use Raspberry Pi and a motor controller to provide the voltage to the drive system of the car. The motor will run faster or slower based on the level of voltage that is applied to these wires and the polarity of the voltage will determine the direction. Raspberry Pi will need to provide a positive or negative 6 volt signal to control the speed and direction of the car.

You'll also need to replace the control signals that go to the front of the car for the steering. This is a bit more difficult. The following is the connector that goes to the front of the car:

The five-pin connector that comes from the control module is shown in the following:

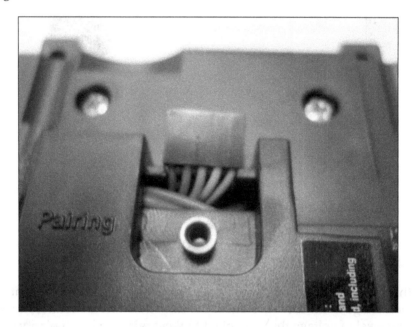

The trick is to determine how the wires control the steering. One way to determine this is by opening up the unit, the following is how it looks from inside:

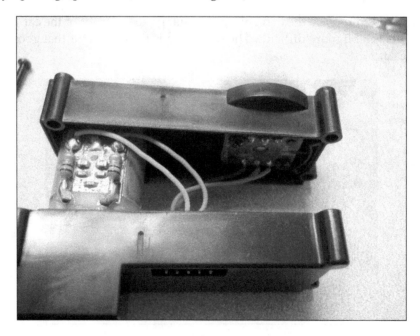

As you can see in the previous image, the blue and yellow wires are attached to a DC motor and the orange, brown, and red wires are attached to another control circuit. The motor will drive the wheels left or right, the polarity of the voltage will determine the direction, and its magnitude will cause the wheels to turn more or less sharply. The orange, brown, and red wires are interesting as their purpose is a bit difficult to discover. To do this, you can hook up a voltmeter and an oscilloscope. The orange and brown wires are straightforward, they are 3.5 volt and GND, respectively. The red wire is a control wire, the signal is a **Pulse Width Modulation (PWM)** signal, a square wave at 330 Hz and 10 percent duty cycle, and it is an enable control signal. Without the signal, the turning mechanism is not engaged.

Now that you understand the signals that are used in the original system to control the car, you can replicate those with Raspberry Pi. To control the steering, Raspberry Pi needs to provide a 3.3 volt DC signal, a GND signal, a 330 Hz, a 3.3 volt PWM signal, and the +/- 6 volt drive signal to the turning mechanism. To make these available, you can use the existing cables, solder some additional cable length, and use some shrink-wrap tubing to create a new connector with the connector that is available in the car:

You'll also need the access to the rear wheel compartment of your car to drive the two rear wheels. The following is how the access will look:

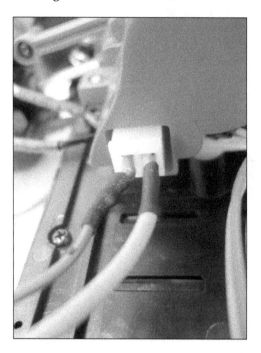

Also, you'll need to connect the battery power to Raspberry Pi, here is the modified connection to get the battery power from the car:

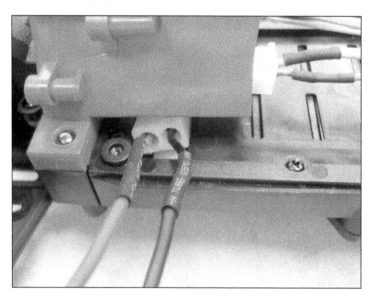

To control the car, you'll need to provide each of the control signals. The +/- 6 volt signals cannot be sourced directly by Raspberry Pi. You'll need some sort of motor controller to source the signal to control the rear wheel drive of the car and turning mechanism of the car. The simplest way to provide these signals is to use a motor shield, an additional piece of hardware that installs at the top of Raspberry Pi and can source the voltage and current to power both of these mechanisms. The RaspiRobot Board V2 is available online and can provide these signals. Here is a picture:

The specifics on the board can be found at http://www.monkmakes.com/?page_id=698. The board will provide two key signals to your RC car, the drive signal and the turn signal. You'll need one more additional signal, the PWM signal that enables the steering control. The following are the steps to connect Raspberry Pi to the board:

1. First, connect the battery power connector to the power connector on the board, as shown in the following:

2. Next, connect the rear drive signal to the motor 1 connectors on the board, similar to the following image:

3. Connect the front drive connector to the motor 2 connectors on the board, as given in the following image:

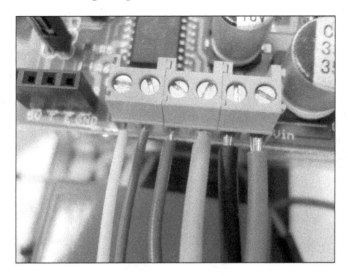

4. Connect the 3.3 volt and GND connectors to the **General Purpose Input/ Output (GPIO)** pins of Raspberry Pi. Here is the layout of these pins:

Pin 1 3.3V	▢ ◯	Pin 2 5V
Pin 3 GPIO2	◯ ◯	Pin 4 5V
Pin 5 GPIO3	◯ ◯	Pin 6 GND
Pin 7 GPIO4	◯ ◯	Pin 8 GPIO14
Pin 9 GND	◯ ◯	Pin 10 GPIO15
Pin 11 GPIO17	◯ ◯	Pin 12 GPIO18
Pin 13 GPIO27	◯ ◯	Pin 14 GND
Pin 15 GPIO22	◯ ◯	Pin 16 GPIO23
Pin 17 3.3V	◯ ◯	Pin 18 GPIO24
Pin 19 GPIO10	◯ ◯	Pin 20 GND
Pin 21 GPIO9	◯ ◯	Pin 22 GPIO25
Pin 23 GPIO11	◯ ◯	Pin 24 GPIO8
Pin 25 GND	◯ ◯	Pin 26 GPIO7
Pin 27 ID_SD	◯ ◯	Pin 28 ID_SC
Pin 29 GPIO5	◯ ◯	Pin 30 GND
Pin 31 GPIO6	◯ ◯	Pin 32 GPIO12
Pin 33 GPIO13	◯ ◯	Pin 34 GND
Pin 35 GPIO19	◯ ◯	Pin 36 GPIO16
Pin 37 GPIO26	◯ ◯	Pin 38 GPIO20
Pin 39 GND	◯ ◯	Pin 40 GPIO21

5. You'll use **Pin 1 3.3V** for the 3.3 volt signal and **Pin 9 GND** for the ground signal. You'll connect one of the GPIO pins so that you can create the 320 Hz, 10 percent duty cycle signal to enable the steering. Connect **Pin 12 GPIO18**, as shown in the following:

Now the hardware is connected.

Controlling the RC car using Raspberry Pi in Python

The hardware is ready, now you can access all this functionality from Raspberry Pi. First, install the library associated with the control board, found at `http://www.monkmakes.com/?page_id=698`. Perform the following steps:

1. Run the command `wget https://github.com/simonmonk/raspirobotboard2/raw/master/python/dist/rrb2-1.1.tar.gz` — this will retrieve the library.

2. Then run `tar -xzf rrb2-1.1.tar.gz` — this will unarchive the files.

3. Type `cd rrb2-1.1` — this will change the directory to the location of the files.

4. Type `sudo python setup.py install` — this will install the files.

Now you'll create some Python code that will allow you to access both the drive motor and the steering motor. The code will look similar to the following:

```
pi@raspberrypi: ~/xmod                                    _  □  X
File Edit Options Buffers Tools Python Help
import RPi.GPIO as GPIO
import time
from rrb2 import *

pwmPin = 18
dc = 10

GPIO.setmode(GPIO.BCM)
GPIO.setup(pwmPin, GPIO.OUT)
pwm = GPIO.PWM(pwmPin, 320)
rr = RRB2()

pwm.start(dc)
rr.set_led1(1)

rr.set_motors(1, 1, 1, 1)

print("Loop, press CTRL C to exit")
while 1:
    time.sleep(0.075)

pwm.stop()
GPIO.cleanup() []

-UU-:**--F1   xmod.py          All L23     (Python)----------------
Auto-saving...done
```

The specifics on the code are as follows:

- `import RPi.GPIO as GPIO`: This will import the `RPi.GPIO` library, allowing you to send out a PWM signal to the front steering mechanism.

- `import time`: This will import the `time` library, allowing you to use the `time.sleep(number_of_milliseconds)`, which causes a fixed delay.

- `from rrb2 import *`: This will import the `rrb2` library, allowing you to control the two DC motors. The `rrb2` is the library you just downloaded from GitHub.

- `pwmPin = 18`: This will set the PWM pin to GPIO Pin 18, which is physically Pin 12 on the Raspberry Pi.

- `dc = 10`: This will set the duty cycle to 10 percent on the PWM signal.

- `GPIO.setmode(GPIO.BCM)`: This will set the definition mode in the `RPi.GPIO` library to the BCM mode, allowing you to specify the physical pin of the PWM signal.

- `GPIO.setup(pwmPin, GPIO.OUT)`: This will set the PWM pin to an output so that you can drive the control circuitry on the steering.

- `pwm = GPIO.PWM(pwmPin, 320)`: This will initialize the PWM signal on the proper pin and set the PWM signal to 320 Hz.

- `rr = RRB2()`: This will instantiate an instance of the motor controller.

- `pwm.start(dc)`: This will start the PWM signal.

- `rr.set_led1(1)`: This will light LED 1 on the motor controller board.

- `rr.set_motors(1, 1, 1, 1)`: This will set both the motors to move so that the vehicle goes in the forward direction. This command will allow you to set the motors to forward or reverse and set it at a specific speed. The first number is the speed of motor one and it goes from 0 to 1. The second numbers is the direction of motor one, 1 is forward and 0 is reverse. The third number is the speed of motor two, which also goes from 0 to 1, and the fourth number is the reverse and forward setting of the second motor, either 1 or 0.

- `print("Loop, press CTRL C to exit")`: This will instruct the user how to stop the program.

- `while 1`: This will keep looping until *Ctrl* + *C* is pressed.

- `time.sleep(0.075)`: Causes the program to wait 0.075 seconds.

- `pwm.stop()`: This will stop the PWM signal.

- `GPIO.cleanup()`: This will cleanup the GPIO driver and prepare for shutdown.

Now you can run the program by typing `sudo python xmod.py`. LED 1 on the control board should turn on, the rear wheels should move in the forward direction, and the steering should turn. This confirms that you have connected everything correctly. To make this a bit more interesting, you can add more dynamic control of the motors by adding some control code. The following is the first part of the python code:

```
pi@raspberrypi: ~/xmod                              _  □  X
File Edit Options Buffers Tools Python Help
import RPi.GPIO as GPIO
import time
from rrb2 import *
import tty
import sys
import termios
def getch():
    fd = sys.stdin.fileno()
    old_settings = termios.tcgetattr(fd)
    tty.setraw(sys.stdin.fileno())
    ch = sys.stdin.read(1)
    termios.tcsetattr(fd, termios.TCSADRAIN, old_settings)
    return ch
pwmPin = 18
dc = 10
GPIO.setmode(GPIO.BCM)
GPIO.setup(pwmPin, GPIO.OUT)
pwm = GPIO.PWM(pwmPin, 320)
rr = RRB2()
pwm.start(dc)
rr.set_led1(1)
var = 'n'
speed1 = 0
speed2 = 0
direction1 = 1
direction2 = 1

while var != 'q':
    var = getch()
    if var == 'l':
-UU-:**--F1   xmodControl.py   Top L1    (Python)----------------------
```

Before you start, you may want to copy your python code in a new file, you can
call it xmodControl.py. In this code you'll have some additional import statements,
which will allow you to sense key presses from the keyboard without hitting the
enter key. This will make the real-time interface seem more real time. The getch()
function senses the actual key press.

The rest of this code will look similar to the previous program. Now the second part of this code is as follows:

```
pi@raspberrypi: ~/xmod
File Edit Options Buffers Tools Python Help
rr.set_led1(1)
var = 'n'
speed1 = 0
speed2 = 0
direction1 = 1
direction2 = 1

while var != 'q':
    var = getch()
    if var == 'l':
        speed1 = 0.5
        direction2 = 1
    if var == 'r':
        speed2 = 0.5
        direction2 = 0
    if var == 's':
        speed2 = 0.1
        direction = 1
    if var == 'f':
        speed1 = 1
        direction1 = 1
    if var == 'b':
        speed1 = 1
        direction1 = 0
    rr.set_motors(speed1, direction1, speed2, direction2)
    time.sleep(0.1)

pwm.stop()
GPIO.cleanup()

-UU-:**--F1   xmodControl.py   Bot L36   (Python)----------------------------
```

The second part of the code is a while loop that takes the input and translates it into commands for your RC car, going forward and backward and turning right and left. This program is quite simple, you'll almost certainly want to add more commands that provide more ways to control the speed and direction.

Accessing the RC car remotely

You can now control your RC Car, but you certainly want to do this without any connected cables. This section will show you how to add a wireless LAN device so that you can control your car remotely. The first step in doing this is to install a Wireless LAN device. There are several possible ways to do this; however, the one that works well, with full documentation, is described at https://learn.adafruit.com/setting-up-a-raspberry-pi-as-a-wifi-access-point/overview.

You should now be able to connect to your Raspberry Pi via the Wireless Access Point. Once you've created the wireless access point, you can login via a VNC connection, this way you can add a USB webcam to your car to make it even easier to control. To do this, first download an application that can support a VNC connection. You can get this on your Raspberry Pi using an application called vncserver. You'll need to install a version of this on your Raspberry Pi by typing `sudo apt-get install tightvncserver` in a terminal window on your Raspberry Pi.

TightVNC Server is an application that will allow you to remotely view your complete graphical desktop. Once you have it installed, you can do the following:

1. You need to start the server by typing `vncserver` in a terminal window on Raspberry Pi.

2. You will be prompted for a password and then asked to verify it, then you will be asked if you'd like to have a view-only password. Remember the password that you have entered, you'll need it to remotely login via a VNC Viewer.

3. You'll need a VNC Viewer application for your remote computer. One choice for Windows users is RealVNC, available at `http://www.realvnc.com/download/viewer/`. When you run it, you will see the following:

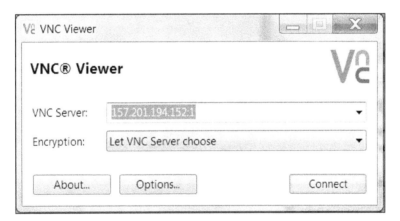

4. Enter the VNC Server address, which is the IP address of your Raspberry Pi, and click on **Connect**. You will get a warning about an unencrypted connection, select **Continue** and you will get the following pop-up window:

5. Type in the password that you entered while starting the vncserver, and you will then get a graphical view of your Raspberry Pi, which looks like the following screenshot:

You can now access all the capabilities of your system; however, they may be slower if you are doing a graphics-intense data transfer. To avoid having to type `vncserver` each time you boot your Raspberry Pi, use the instructions given at `http://www.havetheknowhow.com/Configure-the-server/Run-VNC-on-boot.html`.

Vncserver is also available via Linux. You can use an application called Remote Desktop Viewer to view the remote Raspberry Pi GUI system. If you have not installed this application, install it using the update software application based on the type of Linux system you have. Once you have the software, perform the following steps:

1. Run the application and you will get the following result:

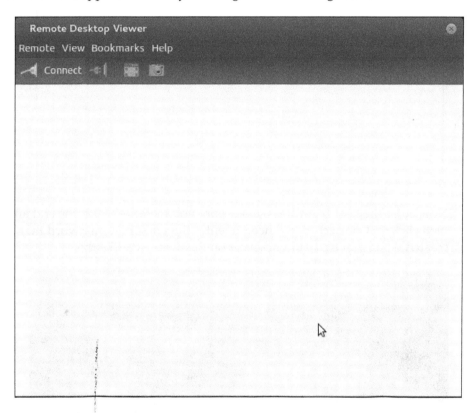

2. Make sure vncserver is running on Raspberry Pi; the easiest way to do this is to log in using SSH and run `vncserver` at the prompt. Now click on **Connect** on **Remote Desktop Viewer**. Fill in the screen as follows, under the **Protocol** selection, choose **VNC**, and you will see the following screenshot:

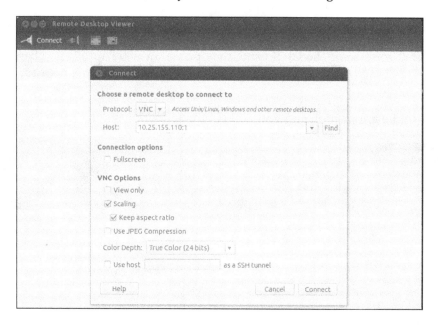

3. Now enter the host IP address, make sure you include a `:1` at the end and then click on **Connect**. You'll need to enter the vncserver password that you created when you first ran vncserver on Raspberry Pi, like this:

You can see the graphical screen of Raspberry Pi. Now you are ready to observe the output of a USB webcam connected to your car. This is quite straightforward, simply plug in a USB webcam and download a video viewer. One such video viewer that works well is luvcview. To install this, type `sudo apt-get install luvcview`.

With all these tools installed, you can now run vncview, bring up a luvcview window so you can see what your RC Car is seeing and control it remotely by running the `xcmodControl.py` program that you wrote earlier. The screen will look similar to the following:

There are a lot of additions that you can make to your Raspberry Pi controlled car, such as adding the joystick control or more autonomy. However, let's move on to the next project.

Summary

Now you know how to work with Raspberry Pi to add its capability to an existing piece of hardware, in this case, an RC Car. In the next chapter, you'll learn how to add Raspberry Pi to a toy robot that can walk, and make it talk and listen to the voice commands.

2
Adding Raspberry Pi to a Humanoid Robot

Modifying an RC car with Raspberry Pi is a wonderful project, but you can take this idea even further by modifying different toys with Raspberry Pi. One class of toys that are excellent candidates for our project are a set of robot toys by **WowWee**. You can purchase these toys from the company directly at http://wowwee.com/, but you can also find used versions of these toys on **eBay** for a significantly lower price.

In this chapter, you'll learn the following:

- How to send and receive voice commands
- How to interpret commands and initiate actions

There are several toys that have excellent possibilities. One such toy is the **WowWee Roboraptor**. The following is an image of this robot:

Another option is the **WowWee Robosapien**. A picture of this robot is given in the following image:

You'll use this robot for your project, as it has more functionality and is easier to modify. Specifically, you're going to connect to the internal serial bus so that you, not the remote, can send commands. You'll be adding **Arduino UNO** to handle the real-time communications between Raspberry Pi and the robot. Here are the steps:

1. First, you'll need to disassemble the robot to get the access to the main controller board. To do this, lay the robot face down, so that you have access to the back. Remove the plate at the back by unscrewing the four screws that hold it in place. Now, at the top of the exposed board, you will see the main connector. The following is a close-up of the connector:

There are only two wires that you are interested in. The first is the black wire, it is the GND for the Robosapien system. The second is the white wire. This is the serial connection that controls the command for the Robosapien.

2. So, you're going to want to connect a wire to the black wire, but you'll want both ends of the black wire to stay connected to the system. To do this, melt a bit of the insulation with a soldering iron and then solder another wire at this point. The following is an image:

3. Now, snip the white wire and connect a wire to the end that is connected to the white header connector, similar to this image:

You may want to add some heat-shrink tubing to cover your connections.

4. Finally, drill a hole in the back shell of the robot so that you can run both of these cables out of the unit, as shown in the following image:

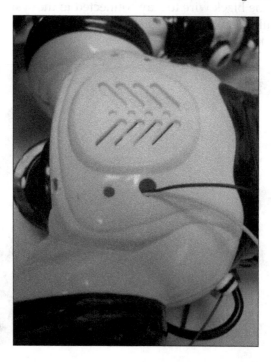

5. You should also drill two more holes on either side of the shell; you can use these to attach Raspberry Pi to the robot with cable ties. Now, you can put the shell back onto the robot.

6. Now you'll connect these two wires to Arduino UNO. Other versions of the Arduino board could also be used for this purpose. The reason you'll need to use Arduino is that the bit patterns that are sent to the robot are at a fairly high rate and need to be created by a processor dedicated to this type of communication. This will allow Raspberry Pi to be involved in other processor-intensive activities, such as speech or object recognition, and yet keep the communication flowing at the right rate.

7. Connect the GND wire to one of the **GND** pins on Arduino. Then, connect the other wire to **Pin 9** on Arduino. These connections will look similar to the following:

8. The final step is to create the code that will send the proper commands to the Arduino board. The code is listed as follows:

```
volatile int viRobsapienCmd = -1;  // A robosapien command
  sent over the UART request
// Some but not all RS commands are defined
#define RSTurnRight        0x80
#define RSRightArmUp       0x81
#define RSRightArmOut      0x82
#define RSTiltBodyRight    0x83
#define RSRightArmDown     0x84
#define RSRightArmIn       0x85
#define RSWalkForward      0x86
#define RSWalkBackward     0x87
#define RSTurnLeft         0x88
#define RSLeftArmUp        0x89
#define RSLeftArmOut       0x8A
#define RSTiltBodyLeft     0x8B
#define RSLeftArmDown      0x8C
#define RSLeftArmIn        0x8D
#define RSStop             0x8E
#define RSWakeUp           0xB1
#define RSBurp             0xC2
#define RSRightHandStrike 0xC0
#define RSNoOp             0xEF
#define RSRightHandSweep  0xC1
#define RSRightHandStrike2 0xC3
#define RSHigh5            0xC4
#define RSFart             0xC7
#define RSLeftHandStrike   0xC8
#define RSLeftHandSweep   0xC9
#define RSWhistle          0xCA
#define RSRoar             0xCE
int LedControl = 13;    // Show when control on
int IROut= 9;           // Where the echoed command will
  be sent from
```

```
int bitTime=516;              // Bit time (Theoretically 833
  but 516)
///////////////////////////////////////////////////////
// Begin Robosapien specific code
///////////////////////////////////////////////////////
// send the command 8 bits
void RSSendCommand(int command) {
  Serial.print("Command: ");
  Serial.println(command, HEX);
  digitalWrite(IROut,LOW);
  delayMicroseconds(8*bitTime);
  for (int i=0;i<8;i++) {
    digitalWrite(IROut,HIGH);
    delayMicroseconds(bitTime);
    if ((command & 128) !=0) delayMicroseconds(3*bitTime);
    digitalWrite(IROut,LOW);
    delayMicroseconds(bitTime);
    command <<= 1;
  }
  digitalWrite(IROut,HIGH);
  delay(250); // Give a 1/4 sec before next
}
// Set up RoboSpapien functionality
void RSSetup()
{
  pinMode(IROut, OUTPUT);
  pinMode(LedControl,OUTPUT);
  digitalWrite(IROut,HIGH);
  RSSendCommand(RSBurp);
}


// Loop for RoboSapien functionality
void RSLoop()
{
  digitalWrite(LedControl,HIGH);
  // Has a new command come?
```

```
    if(viRobsapienCmd != -1)
      {
       RSSendCommand(viRobsapienCmd);
      viRobsapienCmd = -1;
      }
    digitalWrite(LedControl,LOW);
}
void setup()
{
  Serial.begin(9600);
  Serial.println("RobSapien Start");
  RSSetup();
}
void loop()
{
  if (Serial.available() > 0) {
    // read the incoming byte:
    char str = Serial.read();
    switch (str) {
      case 'a':
        viRobsapienCmd = RSTurnRight;
        break;
      case 'b':
        viRobsapienCmd = RSRightArmUp;
        break;
      case 'c':
        viRobsapienCmd = RSRightArmOut;
        break;
      case 'd':
        viRobsapienCmd = RSTiltBodyRight;
        break;
      case 'e':
        viRobsapienCmd = RSRightArmDown;
        break;
      case 'f':
        viRobsapienCmd = RSRightArmIn;
```

```
        break;
case 'g':
  viRobsapienCmd = RSWalkForward;
  break;
case 'h':
  viRobsapienCmd = RSWalkBackward;
  break;
case 'i':
  viRobsapienCmd = RSTurnLeft;
  break;
case 'j':
  viRobsapienCmd = RSLeftArmUp;
  break;
case 'k':
  viRobsapienCmd = RSLeftArmOut;
  break;
case 'l':
  viRobsapienCmd = RSTiltBodyLeft;
  break;
case 'm':
  viRobsapienCmd = RSLeftArmDown;
  break;
case 'n':
  viRobsapienCmd = RSLeftArmIn;
  break;
case 'o':
  viRobsapienCmd = RSStop;
  break;
case 'p':
  viRobsapienCmd = RSWakeUp;
  break;
case 'q':
  viRobsapienCmd = RSBurp;
  break;
case 'r':
  viRobsapienCmd = RSRightHandStrike;
```

```
        break;
      case 's':
        viRobsapienCmd = RSRightHandSweep;
        break;
      case 't':
        viRobsapienCmd = RSRightHandStrike2;
        break;
      case 'u':
        viRobsapienCmd = RSHigh5;
        break;
      case 'v':
        viRobsapienCmd = RSFart;
        break;
      case 'w':
        viRobsapienCmd = RSLeftHandStrike;
        break;
      case 'x':
        viRobsapienCmd = RSRightHandSweep;
        break;
      case 'y':
        viRobsapienCmd = RSLeftHandSweep;
        break;
      case 'z':
        viRobsapienCmd = RSWhistle;
        break;
      case 'A':
        viRobsapienCmd = RSRoar;
        break;
      default:
        viRobsapienCmd = RSNoOp;
    }
  }
  //RS routine
  RSLoop();
}
```

This Arduino code will take an input from the **Serial Monitor**, in this case, a USB connection from Raspberry Pi, and then turns it into the appropriate command for the WowWee robot. Once you have uploaded the code to Arduino, either using an external PC or Raspberry Pi, you can use the Arduino IDE's Serial Monitor capability to send individual letter commands, and the robot should respond to the commands.

 If you are unfamiliar with the Arduino IDE application, Arduino is well documented at https://www.arduino.cc/, including how to upload code and how to use the Serial Monitor to communicate with the Arduino.

Now that the robot works, you can add the following Python program that will send the commands we just saw:

```
pi@raspberrypi: ~/wowee
File Edit Options Buffers Tools Python Help
#!/usr/bin/python

import serial
import sys

ser = serial.Serial('/dev/ttyACM0', 9600, timeout = 1)
total = len(sys.argv)
cmdargs = str(sys.argv)

if total > 1:
    x = sys.argv[1]
    ser.write(x);
    s = ser.read(100);
#    print s




-UU-:----F1  argControl.py    All L1      (Python)--------------------------
For information about GNU Emacs and the GNU system, type C-h C-a.
```

To run this program, type python argControl.py f, and the robot will respond to that command. To make this program executable without the python command, type chmod +x argControl.py, and you will now be able to run the program by typing ./argControl.py f. You'll need this later when you want to run this program from your voice control program.

Giving your robot voice commands

Now that your robot knows how to respond to the commands from the Python program, you can now add the capability to your robot to respond to voice commands. You'll also add the capability to allow your robot to speak, this will make the robot more interactive.

To add these capabilities to your robot, you'll need to add some hardware. This project requires a USB microphone and speaker adapter. Raspberry Pi itself has an audio output but does not have an audio input. So, you'll need the following three pieces of hardware:

- A USB device to plug in a microphone and speaker

- A microphone that can plug into the USB device

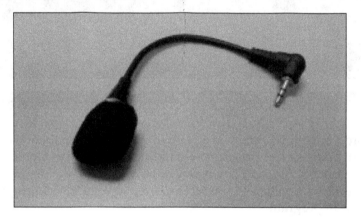

- A powered speaker that can plug into the USB device

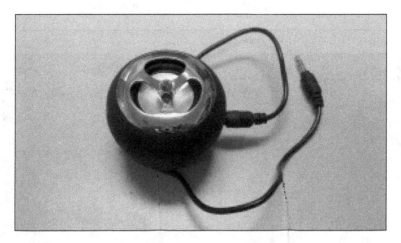

Fortunately, these devices are inexpensive and widely available. Make sure that the speaker is powered because your board will generally not be able to drive a passive speaker with enough power for your applications. The speaker can use either internal battery power or can get its power from a USB connection.

Now, we move on to allowing Raspberry Pi access these devices. You can execute the following instructions in either of the two following ways:

- If you are still connected to the display, keyboard, and mouse, log in to the system and use the GUI by opening an **LXTerminal** window

- If you are only connected through LAN, you can do all this using an SSH terminal window; and as soon as your board indicates that it has power, open up an SSH terminal window using **PuTTY** or any similar terminal emulator

Downloading the example code

You can download the example code files from your account at http://www.packtpub.com for all the Packt Publishing books you have purchased. If you purchased this book elsewhere, you can visit http://www.packtpub.com/support and register to have the files e-mailed directly to you.

Plug the devices into a USB port. Once the terminal window comes up, type
`cat /proc/asound/cards`. You will get the following response:

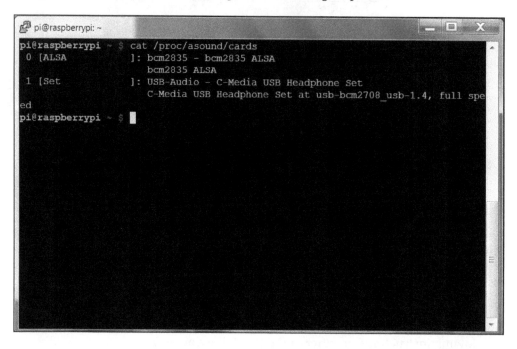

There are two possible audio devices. The first is the internal Raspberry Pi audio that
is connected to the audio port and the second is your USB audio plugin. You could
use the USB audio plugin to record the sound and Raspberry Pi for the audio output
to play the sound. It is easier to just use the USB audio plugin to create and record
sound.

First, you will play some music to test whether the USB sound device is working.
You'll need to configure your system to search for your USB audio plugin and use
it as the default plugin to play and record sound. To do this, you'll need to add a
couple of libraries to your system. The first libraries are the **Advanced Linux Sound
Architecture (ALSA)** libraries. These will enable your sound system on Raspberry Pi
by performing the following steps:

1. Install two libraries that are associated with ALSA by typing `sudo apt-get install alsa-base alsa-utils`

2. Then, install some files that help in providing the sound library by typing `sudo apt-get install libasound2-dev`

You'll use an application named **alsamixer** to control the volume of the input and output of your USB sound card. To do this, perform the following steps:

1. Type `alsamixer` in the Command Prompt. You will see a screen that will be similar to the following screenshot:

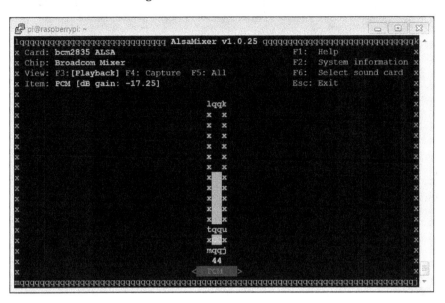

2. Press *F6* and select your USB sound device using the arrow keys. For example, refer to the following screenshot:

`C-Media USB Audio Device` is my USB audio device. You should now be able to see a screen that looks similar to the following screenshot:

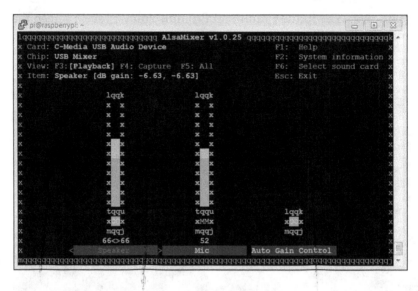

3. You can use the arrow keys to set the volume for both, the speakers and the microphone. Use the *M* key to unmute the microphone. In the preceding screenshot, `MM` is mute and `oo` is unmute.

4. Let's make sure that our system knows about our USB sound device. At the Command Prompt, type `aplay -l`. You should now be able to see the following screenshot:

If this does not work, try `sudo aplay -l`. You are going to add a file to your home directory with the name `.asoundrc`. This will be read by your system and used to set your default configuration. To do this, perform the following steps:

1. Open the file named `.asoundrc` using the editor of your choice.

2. Type `pcm.!default sysdefault:Set`. The `Set` is the variable that appears right after `card 1:` in the output of the `aplay -l` command.

3. Save the file. The file should appear as follows:

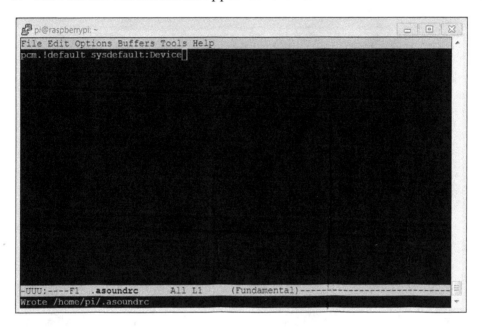

This will tell the system to use your USB device as default. Reboot your system again.

Now you can play some music. To do this, you need a sound file and a device to play it. You can copy a simple .wav file to your Raspberry Pi. If you are using a Linux machine as your host, you can also use the scp command from the command line to transfer the file. You can download some music on Raspberry Pi using a web browser if you have a keyboard, mouse, and display connected. You are going to use the application named **aplay** to play your sound. Type aplay Dance.wav to see whether you can play music using the aplay music player. You will see the result (and hopefully, hear it) as shown in the following screenshot:

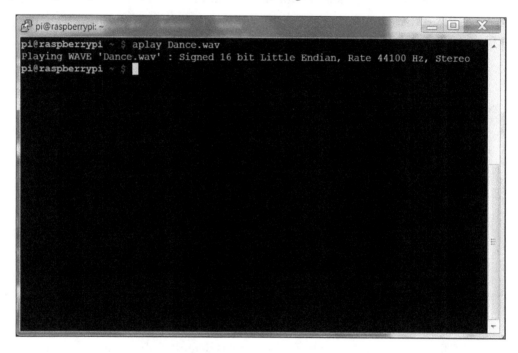

If you don't hear any music, check the volume you set with alsamixer and the power cable of your speaker. Also, aplay can be a bit finicky about the type of files it accepts, so you may be required to try different .wav files until aplay accepts one. One more thing to try, in case the system doesn't seem to know about the program, is to type sudo aplay Dance.wav.

Now that you can play sound, you can also record some sound. To do this, you'll have to use the **arecord** program. At the prompt, type `arecord -d 5 -r 48000 test.wav`. This will record the sound at a sample rate of 48000 Hz per 5 seconds. Once you have typed the command, either speak into the microphone or make some other recognizable sound. You will see the following output on the terminal:

```
pi@raspberrypi: ~

pi@raspberrypi ~ $ aplay Dance.wav
Playing WAVE 'Dance.wav' : Signed 16 bit Little Endian, Rate 44100 Hz, Stereo
pi@raspberrypi ~ $ arecord -d 5 -r 48000 test.wav
Recording WAVE 'test.wav' : Unsigned 8 bit, Rate 48000 Hz, Mono
pi@raspberrypi ~ $
```

Once you have created the file, play it with aplay. Type `aplay test.wav` and you should be able to hear the recording. If you can't hear your recording, check alsamixer to make sure that your speakers and microphone are both unmuted.

Now you can play music or other sound files using your Raspberry Pi. You can change the volume of your speaker and record your voice or other sounds on the system. You're now ready for the next step.

Using eSpeak to allow your robot to respond in voice

Sound is an important tool in our robotic toolkit, but you will want to do more than just play music. Let's make our robot speak. You're going to start with enabling eSpeak, an open source application that provides us with a computer voice. It is a voice generation application. To get this free functionality, download the eSpeak library by typing `sudo apt-get install espeak` at the prompt. The download may take a while but the prompt will reappear when it is complete. Now, let's see if Raspberry Pi has a voice. Type the `espeak "hello"` command. The speaker should emit a hello in a computer generated voice. If it does not, check the speakers and the volume level.

Now that we have a computer generated voice, you may want to customize it. eSpeak offers a fairly complete set of customization features, including a large number of languages, voices, and other options. To access these, you can type in the options at the command-line prompt. For example, type in `espeak -v +f3 "hello"` and you should be able to hear a female voice. You can even add a Scottish accent by typing `espeak -v en-sc +f3 "hello"`.

There are a lot of choices with respect to the voices that you might use with eSpeak. Feel free to play around and choose your favorite voice. Then, edit the default file to set it to this voice. This default file is in the home directory of eSpeak. However, don't expect to get the kind of voices that you hear from computers in the movies; those are actors and not computers. Although, one day, we will hopefully reach a stage where the computers will sound a lot more like real people.

Using pocketsphinx to accept your voice commands

Now that your robot can talk, you'll also want it to obey voice commands. This section will show you how to add speech recognition to your robotic projects. This isn't nearly as simple as the speaking part but, thankfully, you have some significant help from the open source development community. You are going to download a set of capabilities named **pocketsphinx**, which will allow our project to listen to our commands.

The first step is downloading the pocketsphinx capabilities. Unfortunately, this is not quite as user-friendly as the eSpeak process, so follow along the steps carefully. There are two possible ways to do this. If you have a keyboard, mouse, and display connected or want to connect through vncserver, you can do this graphically by performing the following steps:

1. Go to the **Sphinx** website hosted by **Carnegie Mellon University (CMU)** at `http://cmusphinx.sourceforge.net`. This is an open source project that provides you with the speech recognition software. With our smaller embedded system, we will be using the pocketsphinx version of this code.

2. You will need to download two pieces of software modules—**sphinxbase** and pocketsphinx. Select the **DOWNLOAD** option at the top of the page and then find the latest version of both of these packages. Download the `.tar.gz` version of the packages and move them to the `/home/pi` directory of your Raspberry Pi.

Another way to accomplish this is to use **Wget** directly from the command prompt of Raspberry Pi. If you want to do it this way, perform the following steps:

1. To use Wget on your host machine, find the link to the file that you wish to download. In this case, go to the Sphinx website hosted by CMU at `http://cmusphinx.sourceforge.net`. This is an open source project that provides you with the speech recognition software. With your smaller embedded system, you will be using the pocketsphinx version of this code.

2. You will need to download two pieces of software modules, namely sphinxbase and pocketsphinx. Select the **DOWNLOAD** option at the top of the page and then find the latest version of both these packages. Right-click on the `sphinxbase-0.8.tar.gz` file (considering that 0.8 is the latest version) and select **Copy link location**. Now, open a PuTTY window in Raspberry Pi, and after logging in, type `wget` and paste the link that you just copied. This will download the `.tar.gz` version of sphinxbase. Now follow the same procedure with the latest version of pocketsphinx.

Before you build these, you need two libraries. The first library is `libasound2-dev`. Type `sudo apt-get install libasound2-dev`. The second library is called **Bison**. This is a general purpose, open source parser that will be used by pocketsphinx. To get this package, type `sudo apt-get install bison`.

Once everything is downloaded and installed, you can build pocketsphinx. Firstly, your home directory, with the `.tar.gz` files of pocketsphinx and sphinxbase, should look as shown in the following screenshot:

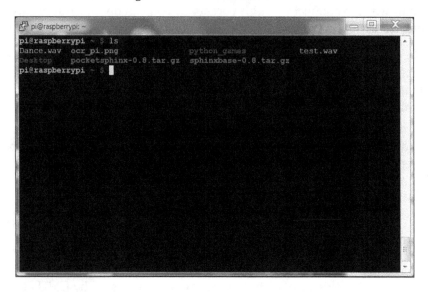

To unpack and build the sphinxbase module, type
`sudo tar -xzvf sphinx-base-0.y.tar.gz`, where y is the version number; in our example, it is 8. This should unpack all the files from the archive into a directory named `sphinxbase-0.8`. Now type `cd sphinxbase-0.8`. Listing of the files should look something similar to the following screenshot:

To build the application, start by issuing the `sudo ./configure --enable-fixed` command. This command will check whether everything is okay with the system and then configure a build.

Now you are ready to actually build the sphinxbase code base. This is a two-step process, which is as follows:

1. Type `sudo make` and the system will build all the executable files.
2. Type `sudo make install` to install all the executables onto the system.

Now, you need to make the second part of the system — the pocketsphinx code. Go to the home directory, and decompress and unarchive the code by typing `tar -xzvf pocketsphinx-0.8.tar.gz`. Now, the files will be unarchived and you can build the code. Installing these files is a three-step process, as follows:

1. Type `cd pocketsphinx-0.8` to go to the pocketsphinx directory, and then type `sudo ./configure` to check whether you are ready to build the files.
2. Type `sudo make` and wait for everything to build.
3. Type `sudo make install`.

> Several possible additions to our library installations will be useful later if you are going to use your pocketsphinx capability with Python as the coding language. You can install **Python-Dev** using `sudo apt-get install python-dev`. Similarly, you can get **Cython** using `sudo apt-get install cython`. You can also choose to install **pkg-config**, a utility that can sometimes help in dealing with complex compiles. Install it using `sudo apt-get install pkg-config`.

Once the installation is complete, you'll need to let the system know where your files are. To do this, use your favorite editor and change the `/etc/ld.so.conf` file by adding a line to the file so it looks as follows:

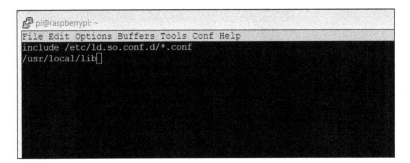

Type `sudo /sbin/ldconfig` and the system will be aware of your pocketsphinx libraries. Now that everything is installed, you can try our speech recognition. Reboot the system and then type `cd /home/pi/pocketsphinx-0.8/src/programs` to go to a directory to try a demo program; then type `./pocketsphinx_continuous`. This program takes input from the microphone and turns it into speech. After running the command, you'll get a lot of irrelevant information, and then you will see the following screenshot:

```
pi@raspberrypi: ~/pocketsphinx-0.8/src/programs
INFO: ngram_model_dmp.c(288):     436879 = LM.bigrams(+trailer) read
INFO: ngram_model_dmp.c(314):     418286 = LM.trigrams read
INFO: ngram_model_dmp.c(339):      37293 = LM.prob2 entries read
INFO: ngram_model_dmp.c(359):      14370 = LM.bo_wt2 entries read
INFO: ngram_model_dmp.c(379):      36094 = LM.prob3 entries read
INFO: ngram_model_dmp.c(407):        854 = LM.tseg_base entries read
INFO: ngram_model_dmp.c(463):       5001 = ascii word strings read
INFO: ngram_search_fwdtree.c(99): 788 unique initial diphones
INFO: ngram_search_fwdtree.c(147): 0 root, 0 non-root channels, 60 single-phone
words
INFO: ngram_search_fwdtree.c(186): Creating search tree
INFO: ngram_search_fwdtree.c(191): before: 0 root, 0 non-root channels, 60 singl
e-phone words
INFO: ngram_search_fwdtree.c(326): after: max nonroot chan increased to 13428
INFO: ngram_search_fwdtree.c(338): after: 457 root, 13300 non-root channels, 26
single-phone words
INFO: ngram_search_fwdflat.c(156): fwdflat: min_ef_width = 4, max_sf_win = 25
INFO: continuous.c(371): /home/pi/pocketsphinx-0.8/src/programs/.libs/lt-pockets
phinx_continuous COMPILED ON: Nov  8 2013, AT: 18:29:54

Warning: Could not find Mic element
Warning: Could not find Capture element
READY....
```

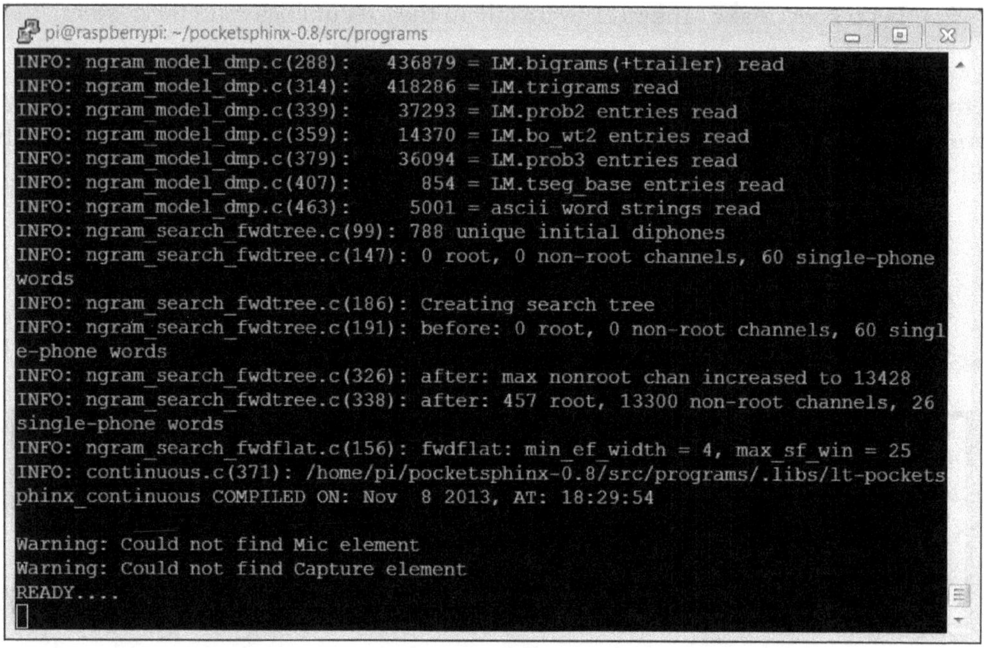

The `INFO` and `Warning` statements come from the C or C++ code and are there for debugging purposes. Initially, they will warn you that they cannot find your `Mic` and `Capture` elements, but when Raspberry Pi finds them, it will print out `READY.....` If you have set things up as described previously, you are ready to give your Raspberry Pi a command. Say hello into the microphone. When it senses that you have stopped speaking, it will process your speech and after giving lots of irrelevant information, it will eventually show the commands, as shown in the following screenshot:

```
pi@raspberrypi: ~                                                    _ □ X
INFO: ngram_model_arpa.c(195): Reading bigrams
INFO: ngram_model_arpa.c(533):         72 = #bigrams created
INFO: ngram_model_arpa.c(534):         14 = #prob2 entries
INFO: ngram_model_arpa.c(542):          7 = #bo_wt2 entries
INFO: ngram_model_arpa.c(292): Reading trigrams
INFO: ngram_model_arpa.c(555):         62 = #trigrams created
INFO: ngram_model_arpa.c(556):          8 = #prob3 entries
INFO: ngram_search_fwdtree.c(99): 40 unique initial diphones
INFO: ngram_search_fwdtree.c(147): 0 root, 0 non-root channels, 12 single-phone
words
INFO: ngram_search_fwdtree.c(186): Creating search tree
INFO: ngram_search_fwdtree.c(191): before: 0 root, 0 non-root channels, 12 singl
e-phone words
INFO: ngram_search_fwdtree.c(326): after: max nonroot chan increased to 194
INFO: ngram_search_fwdtree.c(338): after: 40 root, 66 non-root channels, 11 sing
le-phone words
INFO: ngram_search_fwdflat.c(156): fwdflat: min_ef_width = 4, max_sf_win = 25
INFO: continuous.c(427): /home/pi/pocketsphinx-0.8/src/programs/.libs/lt-pockets
phinx_continuous COMPILED ON: Jun 19 2015, AT: 08:53:39

Warning: Could not find Mic element
Warning: Could not find Capture element
READY....
```

Notice the 000000000: hello command. It recognized your speech! You can try other words and phrases too. The system is very sensitive so it may pick up the background noise. You are also going to find that it is not very accurate. We'll deal with this in a moment. To stop the program, press *Ctrl + C*.

There are two ways to make your voice recognition more accurate. One is to train the system to understand your voice more accurately. This is a bit complex but if you want to know more, go to the pocketsphinx website of CMU.

The second way to improve accuracy is to limit the number of words that your system uses to determine what you are saying. The default has literally thousands of word possibilities, so pocketsphinx may choose the wrong word if the two words are close. To avoid this, you can make your own dictionary to restrict the words pocketsphinx has to choose from. To create your own dictionary, follow the instructions at http://cmusphinx.sourceforge.net/wiki/tutorialdict.

Your system can now understand your voice commands! In the next section of this chapter, you'll learn how to use this input for the project to respond.

Interpreting commands and initiating actions

Now that the system can both hear and speak, you'll want to provide the robot with the capability to respond to your speech and execute some commands based on the speech input. Next, you're going to configure the system to respond to simple commands.

In order to respond, we're going to edit the `continuous.c` code in the `/home/pi/src/programs` directory. We could create our own C file, but this file is already set up in the **makefile** system and is an excellent starting spot. You can save a copy of the current file as `continuous.c.old` so that you can always get back to the starting program if required. Then, you will need to edit the `continuous.c` file. It is very long and a bit complicated. However, you are specifically looking for a section in the code, which is shown in the following screenshot. Look for the comment line `/* Exit if the first word spoken was GOODBYE */` comment line.

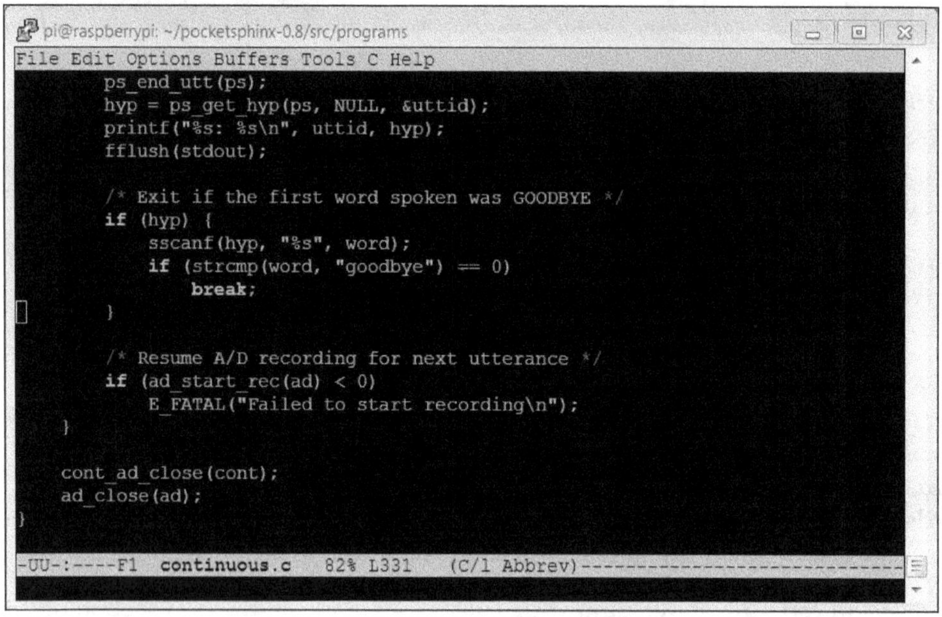

```
pi@raspberrypi: ~/pocketsphinx-0.8/src/programs
File Edit Options Buffers Tools C Help
        ps_end_utt(ps);
        hyp = ps_get_hyp(ps, NULL, &uttid);
        printf("%s: %s\n", uttid, hyp);
        fflush(stdout);

        /* Exit if the first word spoken was GOODBYE */
        if (hyp) {
            sscanf(hyp, "%s", word);
            if (strcmp(word, "goodbye") == 0)
                break;
        }

        /* Resume A/D recording for next utterance */
        if (ad_start_rec(ad) < 0)
            E_FATAL("Failed to start recording\n");
    }

    cont_ad_close(cont);
    ad_close(ad);
}
-UU-:----F1   continuous.c   82% L331   (C/1 Abbrev)-------------------------
```

In this section of the code, the word has already been decoded and is held in the `hyp` variable. You can add the code here in order to make your system do things based on the value associated with the word that we decoded. First, let's try to add the capability to respond to hello and goodbye to see whether we can get the program to respond to these commands. You'll need to make changes to the code in the following manner:

- Find the `/* Exit if the first word spoken was GOODBYE */` comment.

- In the statement `if (strcmp(word, "goodbye") == 0)`, change word to hyp and `goodbye` to `good bye`.

- Insert brackets around the `break;` statement and add the `system("espeak" \"good bye\"");` statement just before the `break;` statement.

- Add the other `else if` statement to the clause by typing `else if (strcmp(hyp, "hello") == 0)`. Add brackets after the `else if` statement and inside the brackets, type `system("espeak" \"hello\"");`.

The file should now look as follows:

```
pi@raspberrypi: ~/pocketsphinx-0.8/src/programs
File Edit Options Buffers Tools C Help
        /* Finish decoding, obtain and print result */
        ps_end_utt(ps);
        hyp = ps_get_hyp(ps, NULL, &uttid);
        printf("%s: %s\n", uttid, hyp);
        fflush(stdout);

        /* Exit if the first word spoken was GOODBYE */
        if (hyp) {
            sscanf(hyp, "%s", word);
            if (strcmp(hyp, "good bye") == 0)
                {
                    system("espeak \"good bye\"");
                    break;
                }
            else if (strcmp(hyp, "hello") == 0)
                {
                    system("espeak \"hello\"");
                }

        }
-UU-:----F1   continuous.c    80% L330    (C/1 Abbrev)--------------------
```

Now you need to rebuild your code. As the makefile system already knows how to build the `pocketsphinx_continuous` program, it will rebuild the application if you make a change to the `continuous.c` file at any point of time. Simply type `sudo make` and the file will compile and create a new version of `pocketsphinx_continuous`. To run your new version, type `./pocketsphinx_continuous`. Make sure that you type the `./` at the start.

If everything is set correctly, saying hello should result in a response of hello from your Raspberry Pi. Saying goodbye should elicit a response of goodbye and also shut down the program. Notice that the system command can be used to run any program that runs with a command line. Now, you can use this program to start and run other programs based on the commands. In this case, you'll want to change the code shown to call your python code to issue the commands to the robot, as shown in the following screenshot:

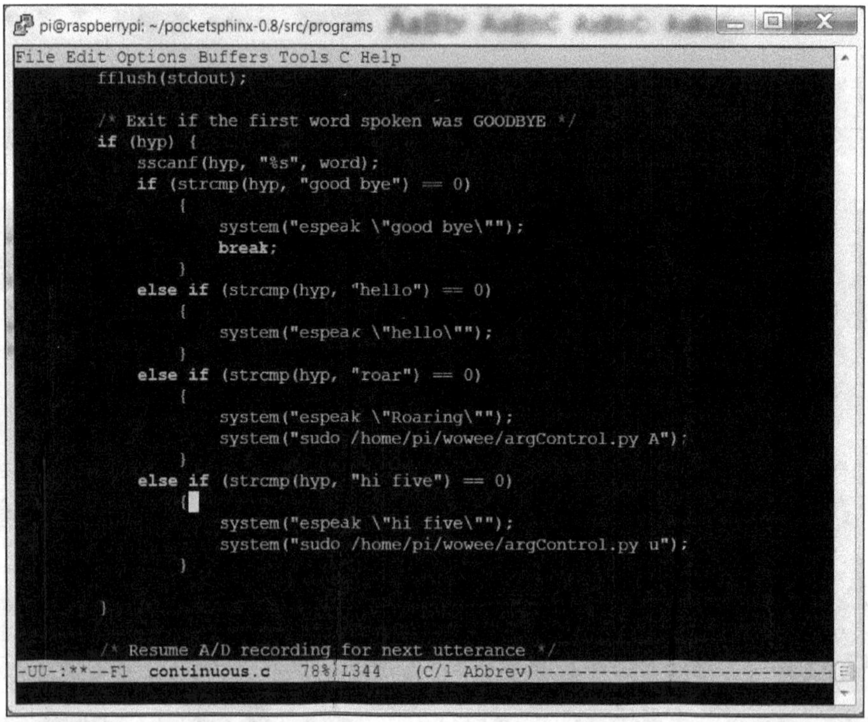

```
pi@raspberrypi: ~/pocketsphinx-0.8/src/programs
File Edit Options Buffers Tools C Help
        fflush(stdout);

        /* Exit if the first word spoken was GOODBYE */
        if (hyp) {
            sscanf(hyp, "%s", word);
            if (strcmp(hyp, "good bye") == 0)
                {
                    system("espeak \"good bye\"");
                    break;
                }
            else if (strcmp(hyp, "hello") == 0)
                {
                    system("espeak \"hello\"");
                }
            else if (strcmp(hyp, "roar") == 0)
                {
                    system("espeak \"Roaring\"");
                    system("sudo /home/pi/wowee/argControl.py A");
                }
            else if (strcmp(hyp, "hi five") == 0)
                {
                    system("espeak \"hi five\"");
                    system("sudo /home/pi/wowee/argControl.py u");
                }

        }

        /* Resume A/D recording for next utterance */
-UU-:**--F1  continuous.c   78% L344   (C/1 Abbrev)--------------------
```

In this case, you'll hook up just two of the many commands that your robot could respond to; you can add the rest of the commands to your `continuous.c` file using this same technique. Now you can give your robot voice commands and it will obey them! Using the directions from the earlier section of this chapter, you can also control you robot remotely using single character commands and add a web cam. You have your very own robotic servant!

Summary

In this chapter, you've learned the basics of how to hack an RC toy car and a toy robot using Raspberry Pi. Feel free to experiment; you can see how easily you can play all sorts of games with your new toys. In the next chapter, you'll learn how to build a robot from the ground up, in this case, a robot that can plan its own path through a set of barriers.

3
Building a Tracked Vehicle
That Can Plan Its Own Path

Now that you are comfortable using Raspberry Pi, the next project you'll tackle is building a tracked robot that can explore and map a room autonomously.

In this chapter, you'll learn the following:

- How to use the **General-purpose input/output (GPIO)** pins to control the speed of a DC motor
- How to control your mobile platform programmatically using Raspberry Pi
- How to connect Raspberry Pi to a USB sonar sensor
- How to connect a digital compass to Raspberry Pi
- How to plan a path for your tracked vehicle

Basic motor control and the tracked vehicle

To build this project, you'll want to start with either a wheeled or a tracked vehicle. There are many options. The following is an image of a tracked platform:

As with the RC car or toy robot, it is difficult to directly connect Raspberry Pi to the DC motors that control the speed and direction of the tracked vehicle. Instead, you'll want to add a DC motor controller for this. You'll use the **RaspiRobot Board V2** for this project, the same board that was introduced in *Chapter 1, Adding Raspberry Pi to an RC Vehicle*.

The board will provide the drive signals for the tracked vehicle. By driving each motor separately, you'll also be able to turn the vehicle. By reversing the signals, you'll be able to change the vehicle's direction and make very sharp turns. The following are the steps to connect the motor control board:

1. Place the motor control board onto the Raspberry Pi.

2. Now, connect the battery power connector to the power connector on the board. Use a battery of 6 to 7 volts, you can either use a 4 AA battery holder or 2S LiPo RC battery. Connect the ground and power wires to the motor control board.

3. Next, connect one of the drive signals to the motor 1 connectors on the board. Connect motor 1 to the right motor and motor 2 to the left.

4. Then, connect the second drive connector to the motor 2 connectors on the board.

The entire set of connections will look like this:

Now you are ready to drive your tracked vehicle using Raspberry Pi.

Controlling the tracked vehicle using Raspberry Pi in Python

The first step to access the functionality is to install the library associated with the control board, which can be found at http://www.monkmakes.com/?page_id=698. You'll create a Python code that will allow you to access the two motors, similar to what you did in the first chapter. The first part of the code, which should look almost the same as the code that you created in the first chapter, will look as follows:

```
pi@raspberrypi: ~/xmod

File Edit Options Buffers Tools Python Help
import RPi.GPIO as GPIO
import time
from rrb2 import *
import tty
import sys
import termios
def getch():
    fd = sys.stdin.fileno()
    old_settings = termios.tcgetattr(fd)
    tty.setraw(sys.stdin.fileno())
    ch = sys.stdin.read(1)
    termios.tcsetattr(fd, termios.TCSADRAIN, old_settings)
    return ch
pwmPin = 18
dc = 10
GPIO.setmode(GPIO.BCM)
GPIO.setup(pwmPin, GPIO.OUT)
pwm = GPIO.PWM(pwmPin, 320)
rr = RRB2()
pwm.start(dc)
rr.set_led1(1)
var = 'n'
speed1 = 0
speed2 = 0
direction1 = 1
direction2 = 1

while var != 'q':
    var = getch()
    if var == 'l':
-UU-:**--F1   xmodControl.py    Top L1      (Python)-----------------------------
```

Now, the second part of the code that will drive the two different motors based on whether you want to go forward, backward, or turn right or left is as follows:

```python
while var != 'q':
    var = getch()
    if var == 'l':
        speed1 = 1
        direction1 = 1
        speed2 = 1
        direction2 = 0
        stop = 1
    if var == 'r':
        speed1 = 1
        direction1 = 0
        speed2 = 1
        direction2 = 1
        stop = 1
    if var == 'f':
        speed1 = 1
        direction1 = 1
        speed2 = 1
        direction2 = 1
        stop = 0
    if var == 'b':
        speed1 = 1
        direction1 = 0
        speed2 = 1
        direction2 = 0
        stop = 0
    if var == 's':
        speed1 = 0
        direction1 = 0
        speed2 = 0
        direction2 = 0
    rr.set_motors(speed1, direction1, speed2, direction2)
    if stop == 1:
        time.sleep(1)
        rr.set_motors(0, 0, 0, 0)
GPIO.cleanup()
```

-UU-:----F1   **track.py**   Bot L40   (Python)---------------------------------

As previously discussed, the `rr.set_motors()` function allows you to specify the speed and direction of each motor independently. Now that you have the basic code to drive your tracked vehicle, you'll need to modify this code so that you can call these functions from another Python program. You'll also need to add some calibrated movement so that your tracked vehicle is able to turn at a certain angle and move forward a set distance. The following is what the code would look like:

```
pi@raspberrypi: ~/tracked                                    _  □  X
File Edit Options Buffers Tools Python Help
import RPi.GPIO as GPIO
import time
from rrb2 import *

rr = RRB2()

def init_vehicle():
    rr.set_led1(1)

def turn_left(angle):
    rr.set_motors(1, 1, 1, 0)
    time.sleep(angle/20)
    rr.set_motors(0, 0, 0, 0)

def turn_right(angle):
    rr.set_motors(1, 0, 1, 1)
    time.sleep(angle/20)
    rr.set_motors(0, 0, 0, 0)

def forward(value):
    rr.set_motors(1, 1, 1, 1)
    time.sleep(value)
    rr.set_motors(0, 0, 0, 0)

def backward(value):
    rr.set_motors(1, 0, 1, 0)
    time.sleep(value)
    rr.set_motors(0, 0, 0, 0)

def stop():
    rr.set_motors(0, 0, 0, 0)

def cleanup():
    GPIO.cleanup()

-UU-:----F1  track.py        All L1      (Python)----------------------
```

The `time.sleep(angle/20)` command in the `turn_right(angle)` and `turn_left(angle)` functions allows the tracked vehicle to move for the right amount of time so that the vehicle moves through the desired angle. You many need to modify this number to get the correct angle of movement. The `time.sleep(value)` command moves the robot for a specific time based on the number stored in `value`. Now that you can move your tracked vehicle, you'll need to connect the sensors to it to know what is going on around the tracked vehicle.

Connecting Raspberry Pi to a USB sonar sensor

One of the easiest ways to sense the presence of objects is to use a sonar sensor. Before adding this capability to your system, here's a little tutorial on sonar sensors. Sonar sensors use ultrasonic sound to calculate the distance from an object. The sound wave travels out from the sensor, as illustrated in the following figure:

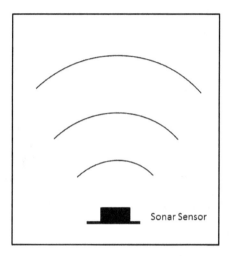

The device emits a sound wave 10 times a second. If an object obstructs these waves, the waves will reflect off of the object and then return to the sensor, as shown in the following figure:

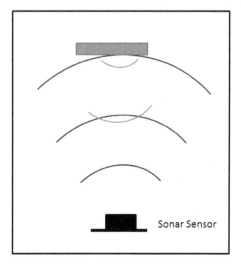

The sensor then measures the returning waves. It uses the time difference between when the sound wave was emitted and when it returned in order to measure its distance from the object.

 Sonar sensors are also quite accurate, normally with a small percentage error, and are not affected by the lighting or color in the environment.

There are several choices if you want to use a sonar sensor to sense the distance. The first option is to use a sonar sensor that connects to the USB port. The following is an image of a USB sonar sensor:

This is the **USB-ProxSonar-EZ** sensor and can be purchased directly from **MaxBotix** or on **Amazon**. There are several models, each has a different distance specification. However, they all work in the same way. There is an inexpensive solution that can be connected to the GPIO of the Raspberry Pi. The following is an image of this sort of inexpensive sonar sensor:

This sensor is less expensive and easy to use. Although it takes a bit of processing power to coordinate the efforts of timing to send and receive signals, the **Raspberry Pi 2 Model B** has the processing power needed.

There are two ways to connect the sensor to Raspberry Pi. You can connect it directly to the motor controller board, as discussed earlier in this chapter. If you are going to do that, there is special connector for the sonar sensor. Here is an image:

To use this connector, simply connect the **Vcc** pin on the sensor to the **5V** pin on the board, the **Trig** pin on the sensor to the **T** pin on the board, the **Echo** pin on the sensor to the **E** pin on the board, and the **Gnd** pin on the sensor to the **GND** pin on the board. You can then use the library for the motor controller board and simply call the `rr.get_distance()` function.

The following are the steps to set up this sonar sensor to sense the distance in case you want to connect it directly to Raspberry Pi's GPIO:

1. Connect it to the GPIO pins on the Raspberry Pi. The first step is to understand the GPIO pins for the Raspberry Pi 2 Model B. Here is a diagram of the layout of the pins:

Pin 1 3.3V		Pin 2 5V
Pin 3 GPIO2		Pin 4 5V
Pin 5 GPIO3		Pin 6 GND
Pin 7 GPIO4		Pin 8 GPIO14
Pin 9 GND		Pin 10 GPIO15
Pin 11 GPIO17		Pin 12 GPIO18
Pin 13 GPIO27		Pin 14 GND
Pin 15 GPIO22		Pin 16 GPIO23
Pin 17 3.3V		Pin 18 GPIO24
Pin 19 GPIO10		Pin 20 GND
Pin 21 GPIO9		Pin 22 GPIO25
Pin 23 GPIO11		Pin 24 GPIO8
Pin 25 GND		Pin 26 GPIO7
Pin 27 ID_SD		Pin 28 ID_SC
Pin 29 GPIO5		Pin 30 GND
Pin 31 GPIO6		Pin 32 GPIO12
Pin 33 GPIO13		Pin 34 GND
Pin 35 GPIO19		Pin 36 GPIO16
Pin 37 GPIO26		Pin 38 GPIO20
Pin 39 GND		Pin 40 GPIO21

In this case, you'll need to connect to the 5 volt connection of the Raspberry Pi pin 2. You also need to connect to the GND, which is pin 6 on Raspberry Pi. Pin 16 is used as an output trigger pin and Pin 18 (GPIO24) as an input to time the echo from the sonar sensor.

2. Now that you know the pins you have to connect to, you'll connect the sonar sensor. There is a problem, as you can't connect the 5 volt return from the sonar sensor directly to the Raspberry Pi GPIO pins, they want 3.3 volts. You need to build a voltage divider that will reduce the 5 volts to 3.3 volts. This can be done with two resistors, which are connected as shown in the following diagram:

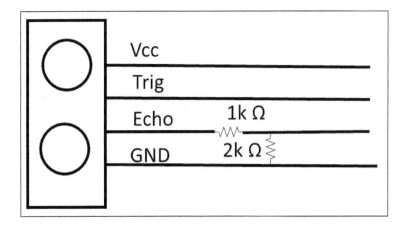

For more information on how the voltage divider works in this configuration, refer to http://www.modmypi.com/blog/hc-sr04-ultrasonic-range-sensor-on-the-raspberry-pi. The combination of these two resistors will reduce the voltage to the desired levels. You may want to put all of this on a small breadboard, as shown in the following:

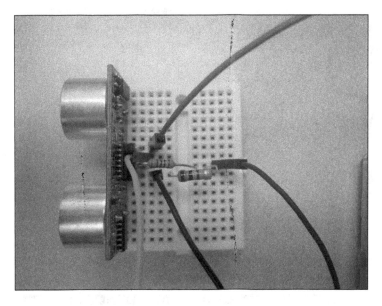

Finally, connect the sensor to the Raspberry Pi, like this:

3. Now that the device is connected, you'll need a bit of code to read in the value; make sure it is settled (a stable measurement), and then convert it to distance. Here is the Python code for this program:

```
pi@raspberrypi: ~
File Edit Options Buffers Tools Python Help
import RPi.GPIO as GPIO
import time
GPIO.setmode(GPIO.BCM)

trig_pin = 23
echo_pin = 24
GPIO.setup(trig_pin,GPIO.OUT)
GPIO.setup(echo_pin,GPIO.IN)

GPIO.output(trig_pin, False)
print "Waiting to settle"
time.sleep(1)
GPIO.output(trig_pin, True)
time.sleep(0.00001)
GPIO.output(trig_pin, False)

while GPIO.input(echo_pin)==0:
    start = time.time()

while GPIO.input(echo_pin)==1:
    end = time.time()

duration = end - start
distance = duration * 17150
distance = round(distance, 2)
print "Distance:",distance,"cm"
GPIO.cleanup()



-UU-:----F1  sonar_sensor.py    All L1    (Python)------------------------------
For information about GNU Emacs and the GNU system, type C-h C-a.
```

4. Now you should be able to run the program and get a result, as shown in the following screenshot:

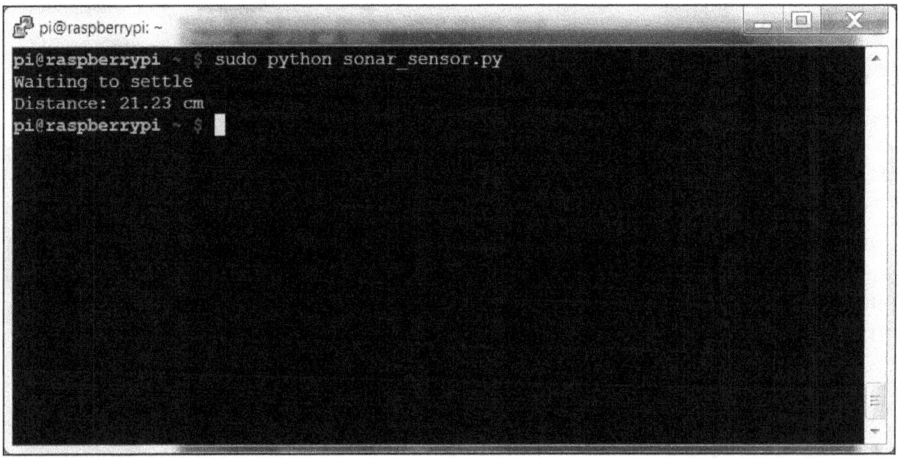
pi@raspberrypi: ~
pi@raspberrypi ~ $ sudo python sonar_sensor.py
Waiting to settle
Distance: 21.23 cm
pi@raspberrypi ~ $

Using this type of sonar sensor provides an inexpensive way to find objects. The final step is to turn this code into a library so that you can call it from the main vehicle program, as shown in the following screenshot:

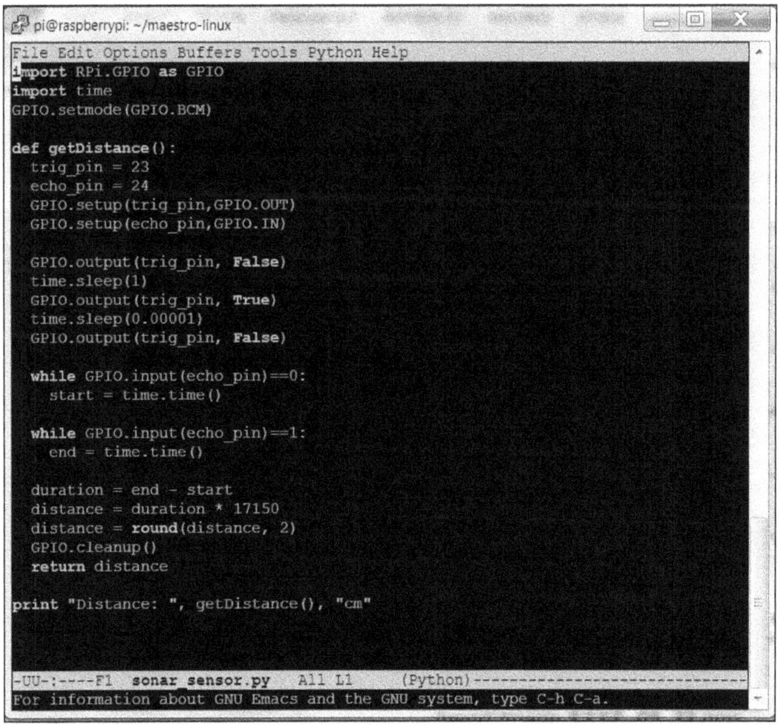
pi@raspberrypi: ~/maestro-linux
File Edit Options Buffers Tools Python Help
import RPi.GPIO as GPIO
import time
GPIO.setmode(GPIO.BCM)
def getDistance():
trig_pin = 23
echo_pin = 24
GPIO.setup(trig_pin,GPIO.OUT)
GPIO.setup(echo_pin,GPIO.IN)
GPIO.output(trig_pin, False)
time.sleep(1)
GPIO.output(trig_pin, True)
time.sleep(0.00001)
GPIO.output(trig_pin, False)
while GPIO.input(echo_pin)==0:
start = time.time()
while GPIO.input(echo_pin)==1:
end = time.time()
duration = end - start
distance = duration * 17150
distance = round(distance, 2)
GPIO.cleanup()
return distance
print "Distance: ", getDistance(), "cm"
-UU-:----F1 sonar_sensor.py All L1 (Python)
For information about GNU Emacs and the GNU system, type C-h C-a.

You can now use this functionality to detect objects when you do your path planning capabilities later in the chapter.

Connecting a digital compass to the Raspberry Pi

One of the important pieces of information that might be useful for your robot, if it is going to plan its own path, is its direction of travel. So, let's learn how to hook up a digital compass to the Raspberry Pi.

There are several chips that provide digital compass capability, one of the most common is the **HMC5883L**, a *3-axis* digital compass chip. This chip is packaged onto a module by several companies but almost all of them result in a similar interface. Here is a picture of one the **GY-271 HMC5883L** triple axis compass magnetometer sensor module available at a number of online retailers:

This type of digital compass uses magnetic sensors to measure the earth's magnetic field. The output of these sensors is then made accessible to the outside world through a set of registers that allow the user to set things like the sample rate and continuous or single sampling. The X, Y, and Z directions are the output using registers as well.

The connections to this chip are straightforward: the device communicates with Raspberry Pi using the **I2C** bus. If you are using the motor controller, then you can connect the device to the I2C bus on the controller board, and if you are using the motor controller to connect to the LIDAR, you'll need to connect the I2C bus to the GPIO pins on Raspberry Pi. At the back of the module, the connections are labelled as shown in the following image:

You then connect the module to the GPIO pins on Raspberry Pi. You need to connect the **VCC** pin on the module to **Pin 1 3.3 V** on Raspberry Pi and **GND** to **Pin 9 GND**. Then connect **SCL** on the module to **Pin 5 GPIO3** and **SDA** to **Pin 3 GPIO2** on the Raspberry Pi. Note that you will not connect the **DRDY (Data Ready)** pin. Now, you are ready to communicate with the device.

Accessing the compass programmatically

Now that the device is connected, you'll need to configure access via the software. Following are the steps:

1. In order to access the compass capability, you'll need to enable the I2C library on Raspberry. The first step in enabling the **analog-to-digital converter (ADC)** is to enable the I2C interface. The I2C interface is a synchronous serial interface and provides more performance than an asynchronous Rx/Tx serial interface. The SCL data line provides a clock while the data flows on the SDA line. The bus also provides addressing so that more than one device can be connected to the master device at the same time. To enable this bus, run `sudo raspi-config` and select the **8 Advanced Options**, as follows:

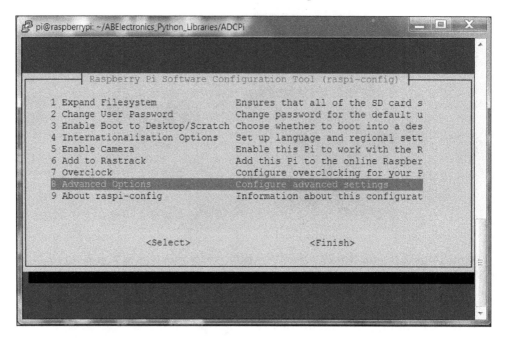

2. Then go to the **A7 I2C** selection and enable the I2C, as shown in the following:

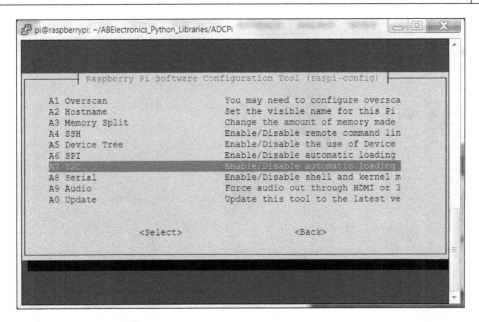

3. Make all the selections to enable the I2C interface and load the library, and then reboot Raspberry Pi.

4. You'll also need to edit the `/etc/modules` file and add the following two lines:

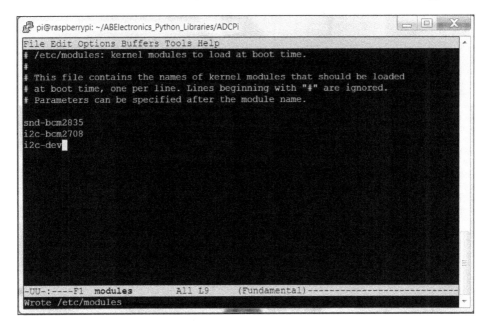

5. Reboot Raspberry Pi. Install the I2C toolkit by typing `sudo apt-get install i2c-tools`. You can see whether I2C is enabled by typing `sudo i2cdetect -y 1`, and then you should see something like this:

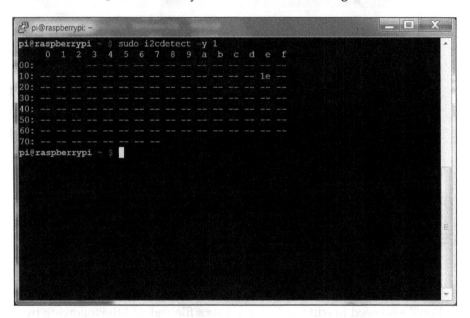

You can see the device at `1e`. Now you can communicate with your digital compass. Here are the steps:

1. You'll need to create a Python program. Before you create your Python code, you need to install the **SMBus** capability to access I2C. This can be done by typing `sudo apt-get install python-smbus`.

2. Now reboot Raspberry Pi and create the following Python code:

```
pi@raspberrypi: ~/tracked

File Edit Options Buffers Tools Python Help
#!/usr/bin/python
import smbus
import time
import math
bus = smbus.SMBus(1)
address = 0x1e

def read_byte(adr):
    return bus.read_byte_data(address, adr)
def read_word(adr):
    high = bus.read_byte_data(address, adr)
    low = bus.read_byte_data(address, adr+1)
    val = (high << 8) + low
    return val
def read_word_2c(adr):
    val = read_word(adr)
    if (val >= 0x8000):
        return -((65535 - val) + 1)
    else:
        return val
def write_byte(adr, value):
    bus.write_byte_data(address, adr, value)

write_byte(0, 0b01110000) # Set to 8 samples @ 15Hz
write_byte(1, 0b00100000) # 1.3 gain LSb / Gauss 1090 (default)
write_byte(2, 0b00000000) # Continuous sampling
scale = 0.92
x_out = read_word_2c(3) * scale
y_out = read_word_2c(7) * scale
z_out = read_word_2c(5) * scale
bearing  = math.atan2(y_out, x_out)
if (bearing < 0):
    bearing += 2 * math.pi
print "Bearing: ", math.degrees(bearing)


-UU-:**--F1  newCompass.py  All L28   (Python)---
```

3. Run the code by typing `python compass.py`, and you should see the following:

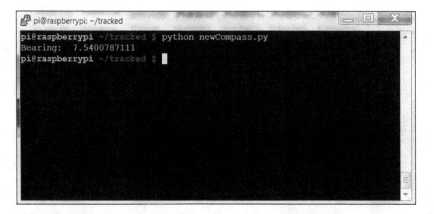

4. The last step is to create a python file and function that can be imported into your tracked vehicle python program. Here is the python code:

```
pi@raspberrypi: ~/tracked
File Edit Options Buffers Tools Python Help
#!/usr/bin/python
import smbus
import time
import math
bus = smbus.SMBus(1)
address = 0x1e

def read_byte(adr):
    return bus.read_byte_data(address, adr)
def read_word(adr):
    high = bus.read_byte_data(address, adr)
    low = bus.read_byte_data(address, adr+1)
    val = (high << 8) + low
    return val
def read_word_2c(adr):
    val = read_word(adr)
    if (val >= 0x8000):
        return -((65535 - val) + 1)
    else:
        return val
def write_byte(adr, value):
    bus.write_byte_data(address, adr, value)

def readDirection():
    write_byte(0, 0b01110000) # Set to 8 samples @ 15Hz
    write_byte(1, 0b00100000) # 1.3 gain LSb / Gauss 1090 (default)
    write_byte(2, 0b00000000) # Continuous sampling
    scale = 0.92
    x_out = read_word_2c(3) * scale
    y_out = read_word_2c(7) * scale
    z_out = read_word_2c(5) * scale
    bearing = math.atan2(y_out, x_out)
    if (bearing < 0):
        bearing += 2 * math.pi
    print "Bearing: ", math.degrees(bearing)
    return math.degrees(bearing)

-UU-:----F1  libCompass.py   All L1     (Python)---------------------
```

Now you can add direction to your project! As you move the device around, you can query the bearing value to see the direction of your tracked vehicle. Make sure that you position your compass away from your electronics, otherwise their magnetic fields may cause distortion on your compass measurements.

The program we just saw is a basic program. If you go to `http://think-bowl.com/raspberry-pi/i2c-python-library-3-axis-digital-compass-hmc5883l-with-the-raspberry-pi/`, you can find more about the other features that are available with this library. Now, you can add the compass capability with just a few lines of code to your tracked robot control program.

Dynamic path planning for your robot

Now that you can see the barriers and also know the direction, you'll want to do dynamic path planning. Dynamic path planning simply means not having the knowledge of all the possible barriers before encountering them. Your robot will have to decide how to proceed while it is in motion. This can be a complex topic but there are some basic concepts that you can understand and apply as you instruct your robot to move around in its surrounding. Let's first address the problem in which you know where you want to go and need to execute a path without barriers and then add the barriers to the path.

Basic path planning

In order to learn about the dynamic path planning, which is planning a path with potential unknown barriers, you need a framework to understand where your robot is and to determine the location of the goal. One of the common framework is an x-y grid. The following is the diagram of such a grid:

						Goal Point 6. 4
			Robot 3. 1			
Reference Point 0. 0						

There are three key points on this grid that you'll need to understand, here is an explanation of these:

1. The lower left point is a fixed reference position. The directions x and y are also fixed and all other positions will be measured with respect to this position and these directions. Each unit is measured with respect to how far the unit travels in time in a single unit.

2. Another important point is the starting location of your robot. Your robot will then keep a track of its location using its x and y coordinates, the position with respect to some fixed reference position in the x direction, or the position with respect to some fixed reference position in the y direction to the goal. It will use the compass to keep track of these directions.

3. The third important point is the position of the goal, also given in x and y coordinates with respect to the fixed reference position. If you know the starting location and angle of your robot, you can plan an optimum (shortest distance) path to this goal. To do this, you can use the goal location, robot location, and some fairly simple math to calculate the distance and angle from the robot to the goal.

To calculate the distance, use the given equation:

$$d = \sqrt{\left(\left(Xgoal - Xgoal \right)^2 + \left(Ygoal - Yrobot \right)^2 \right)}$$

You'll use this equation to tell your robot how far to travel to reach the goal. A second equation will tell your robot the angle at which it needs to travel:

$$\theta = \arctan\left(\frac{Ygoal - Yrobot}{Xgoal - Xrobot} \right)$$

Here is the graphical representation of these two pieces of information:

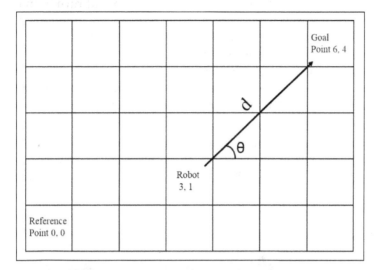

Now that you have a goal angle and distance, you can program your robot to move. To do this, you will write a program to do the path planning and call the movement functions you created earlier in this chapter. You will need to know the distance that your robot travels in a span of time so that you can tell your robot in time units, and not distance units, how far to travel.

You'll also need to be able to translate the distance that might be covered by your robot in a turn; however, this distance may be so small as to be of no importance. If you then know the angle and distance, you can move your robot to the goal.

Here are the steps that you will program:

1. Calculate the distance in units that your robot will travel to reach the goal. Convert this into a number of steps to achieve this distance.

2. Calculate the angle at which your robot will need to travel to reach the goal. You'll use the compass and your robot's turn functions to turn and achieve this angle.

3. Now call the step functions for a required number of times to move your robot the correct distance.

That's it. Now we will use a very simple python code that executes the steps we just saw, using the functions to move the robot forward and turn it. In this case, it makes sense to create a file called `robotLib.py` with all the functions that do the actual servo settings to move the biped robot forward and turn the robot. You'll then import these functions using the `from robotLib import *` statement and your python program can call these functions. This makes the path planning Python program much smaller and more manageable. You'll do the same thing with the compass program, using the `from compass import *` command.

For more information on how to import the functions from one python file to another, refer to http://www.tutorialspoint.com/python/python_modules.htm

Here is a screenshot of the program:

```
pi@raspberrypi: ~/tracked                                          _  □  X
File Edit Options Buffers Tools Python Help
#!/usr/bin/python
import time
from track import *
import math

xpos_robot = int(raw_input("Robot X Position: "))
ypos_robot = int(raw_input("Robot Y Position: "))
xpos_goal = int(raw_input("Goal X Position: "))
ypos_goal = int(raw_input("Goal Y Position: "))

distance = math.sqrt((xpos_goal - ypos_robot)**2 + (ypos_goal - ypos_robot)**2)
angle = round(math.degrees(math.atan2((ypos_goal - ypos_robot), (xpos_goal - xpos_robot))))
if angle < 0:
    angle = angle + 360
print (angle)
# Turn to the right bearing
if (angle) < 180:
    turn_right(angle)
else:
    turn_left(angle)
print (distance)
forward(distance)



-UU-:----F1  robotGoal.py   All L1     (Python)------------------------------
For information about GNU Emacs and the GNU system, type C-h C-a.
```

In this program, the user enters the goal location and the robot decides the shortest direction to the desired location by reading the angle. To make it simple, the robot is placed in the grid, heading in the direction of angle of 0 degrees. If the goal angle is less than 180 degrees, the robot will turn right. If it is greater than 180 degrees, the robot will turn left. The robot turns until the desired angle and its measured angle are within a few degrees. Then, the robot takes the number of steps to reach the goal.

Avoiding obstacles

Planning paths without obstacles, as has been shown, is quite easy. However, it becomes a bit more challenging when your robot needs to walk around the obstacles. Let's look at the case where there is an obstacle in the path that you calculated previously. It might look similar to the following diagram:

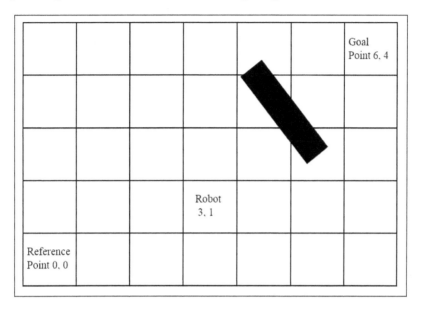

You can still use the same path planning algorithm to find the starting angle; however, you'll now need to use your sonar sensor to detect the obstacle. When your sonar sensor detects the obstacle, you'll need to stop and recalculate the path to avoid the barrier, and also recalculate the desired path to the goal. One very simple way to do this is, when your robot senses a barrier, to turn right at 90 degrees, move a fixed distance, and then recalculate the optimum path. When you turn back to move towards the target, if you sense no barrier, you will move along the optimum path.

However, if your robot encounters the obstacle again, it will repeat the process until it reaches the goal. In this case, using these rules, the robot will travel the following path:

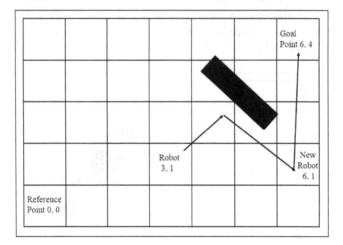

To sense the barrier, you'll use the library calls to the sensor. You can add more accuracy to this robot using the compass to determine your angle. You can do that by importing the compass capability using `from compass import *`. You can also use the `time` library and the `time.sleep` command to add delay among the different statements in the code. You'll need to change your `track.py` library so that the commands don't have a fixed ending time, as follows:

```
import RPi.GPIO as GPIO
import time
from rrb2 import *

rr = RRB2()

def init_vehicle():
    rr.set_led1(1)

def turn_left():
    rr.set_motors(1, 1, 1, 0)

def turn_right():
    rr.set_motors(1, 0, 1, 1)

def forward():
    rr.set_motors(1, 1, 1, 1)

def backward():
    rr.set_motors(1, 0, 1, 0)

def stop():
    rr.set_motors(0, 0, 0, 0)

def cleanup():
    GPIO.cleanup()
```

The following is the first part of this code, the two functions that provide the capability to turn to a known angle using the compass, and a function to calculate the distance and angle and to turn the tracked vehicle at that angle:

```python
pi@raspberrypi: ~/tracked
File Edit Options Buffers Tools Python Help
#!/usr/bin/python
import serial
import time
from track import *
from libCompass import *
from rrb2 import *
import math

def move_angle(angle):
    if angle < 0:
        angle = angle + 360
    bearing = readDirection()
    move_angle = bearing - angle
    if move_angle > 180:
        turn_right()
    elif move_angle < -180:
        turn_left()
    elif (move_angle) < 180 and move_angle > 0:
        turn_left()
    elif move_angle > -180 and move_angle < 0:
        turn_right()
    while(abs(angle - bearing)) > 5:
        time.sleep(.2)
        print abs(angle-bearing)
        bearing = readDirection()
    stop()
#    print "angle", bearing

def positionRobot(xpos, ypos, xpos_goal, ypos_goal):
    print xpos, ypos, xpos_goal, ypos_goal
    distance = math.sqrt((xpos_goal - ypos_robot)**2 + (ypos_goal - ypos_robot)\
**2)
    angle = round(math.degrees(math.atan2((ypos_goal - ypos_robot), (xpos_goal \
- xpos_robot))))
    print "angle",angle
    move_angle(angle)
    print distance
    return distance, angle

xpos robot = int(raw_input("Robot X Position: "))
-UU-:----F1  robotBarrier.py   Top L1    (Python)---------------------------
For information about GNU Emacs and the GNU system, type C-h C-a.
```

The second part of this code shows the main loop. The user enters the robot's current position and desired end position in x and y coordinates. The code calculates the angle and distance and starts the robot on its way. If a barrier is sensed, the unit turns 90 degrees, goes one units distance, and then recalculates the path to the end goal:

```python
pi@raspberrypi: ~/tracked
File Edit Options Buffers Tools Python Help
xpos_robot = int(raw_input("Robot X Position: "))
ypos_robot = int(raw_input("Robot Y Position: "))
xpos_goal = int(raw_input("Goal X Position: "))
ypos_goal = int(raw_input("Goal Y Position: "))

distance, angle = positionRobot(xpos_robot, ypos_robot, xpos_goal, ypos_goal)

start_time = time.time()
forward()
barrier = rr.get_distance()
elapsed_time = 0

while barrier > 10 and elapsed_time < distance:
    elapsed_time = time.time() - start_time
    barrier = rr.get_distance()
    if barrier > 0 and barrier < 10:
        print "barrier", barrier
        distance_traveled = elapsed_time
        new_distance = 1
        ypos_robot = ypos_robot + distance_traveled * math.sin(math.radians(angle))
        ypos_goal_barrier = ypos_robot + new_distance * math.sin(math.radians(angle + 90))
        xpos_robot = xpos_robot + distance_traveled * math.cos(math.radians(angle))
        xpos_goal_barrier = xpos_robot + new_distance * math.cos(math.radians(angle + 90))
        distance = positionRobot(xpos_robot, ypos_robot, xpos_goal_barrier, ypos_goal_barrier)
        start_time = time.time()
        forward()
        elapsed_time = 0
        while elapsed_time < new_distance:
            elapsed_time = time.time() - start_time
        print "Done moving around barrier"
        ypos_robot = ypos_goal_barrier
        xpos_robot = xpos_goal_barrier
        distance = positionRobot(xpos_robot, ypos_robot, xpos_goal, ypos_goal)
        start_time = time.time()
        forward()
        barrier = rr.get_distance()
        elapsed_time = 0
stop()
print "Goal Reached"
-UU-:----F1  robotBarrier.py   40% L57   (Python)------------------------------
```

Now, this algorithm is quite simple, but there are others that have much more complex responses to barriers. You can also see that by adding sonar sensors to the sides, your robot can actually sense when the barrier has ended. You could also provide more complex decision processes about which way to turn to avoid an object. Again, there are many different path finding algorithms. Refer to http://www.academia.edu/837604/A_Simple_Local_Path_Planning_Algorithm_for_Autonomous_Mobile_Robots for an example of this. These more complex algorithms can be explored using the basic functionality that you have built in this chapter.

Summary

You've now added path planning to your tracked robot's capability. Your tracked robot can not only move from point A to point B, but can also avoid barriers that might be in the way. In the next chapter, you'll learn how to build a wheeled robot that can play laser tag.

4
Building a Robot That Can Play Laser Tag

In the previous chapters, you've modified an RC car to control it remotely using Raspberry Pi, you've modified a toy robot to respond to your voice commands and you've also built a tracked vehicle that uses sensors to avoid the barriers and arrive at a desired location. In this chapter, you'll leverage some of these capabilities and then add other capabilities so that you can build a pair of wheeled robots to play laser tag.

In this chapter you'll learn the following:

- Construct a simple two-wheeled platform
- Leverage the wireless LAN interface and a USB webcam to control your robot via a remote computer
- Add a joystick to your host computer
- Connect Raspberry Pi to a laser source
- Connect Raspberry Pi to a laser receiver
- Send and receive laser signals programmatically to enable the laser tag capabilities of your robot

Building and controlling a basic wheeled vehicle

To build this project, you'll want to start with a simple wheeled vehicle. There are many possibilities. The following is a two wheeled vehicle available at many online retail outlets like `https://www.amazon.com` or `http://www.ebay.com`:

First, you'll build the vehicle using the instructions that come with it. The vehicle uses two DC motors, so you'll control the direction and speed of your robot using a DC motor controller. Since it is so flexible and you are already familiar with it, you'll use the RaspiRobot Board V2. The following is an image of the board:

The specifics on the board can be found at http://www.monkmakes.com/?page_
id=698. Connections to the board are very similar to the tracked vehicle connections
that were described in *Chapter 3, Building a Tracked Vehicle That Can Plan Its Own
Path*. You'll place the motor controller on top of the vehicle, connect the battery to the
motor controller, and then connect both the motors, as shown in the following:

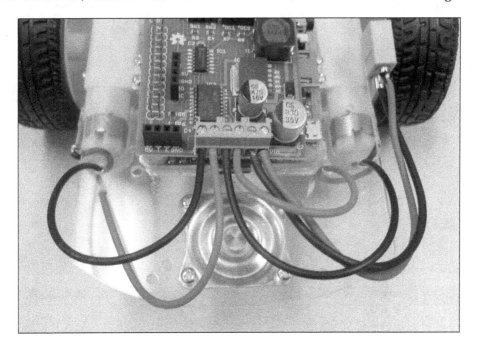

Two of each pair of the red and black connectors go to each one of the motors. The other red and black connectors come from the battery. Once the motors are connected, you are ready to start controlling the speed and direction of your wheeled robot.

Using the robot library to programmatically control your robot

First, install the libraries to support the motor control board, described in the second chapter. Since this robot uses the same control as the wheeled vehicle, you'll use the same code for the simplified library that you used as a library in the second chapter, as shown here:

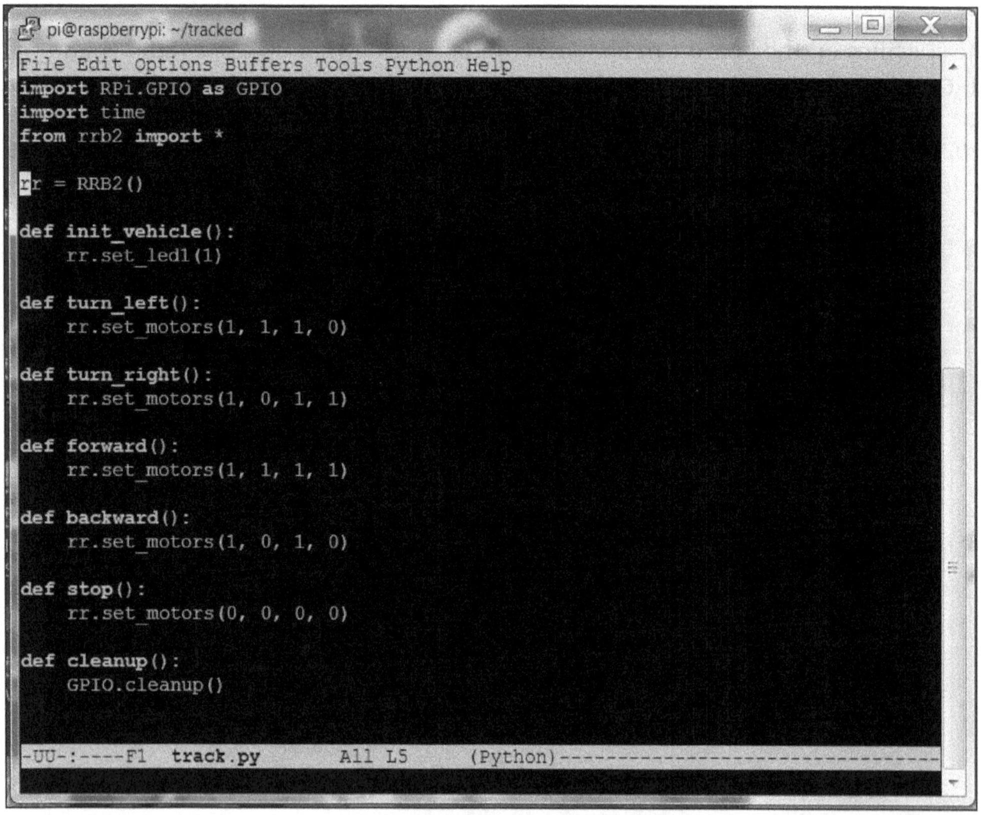

```
pi@raspberrypi: ~/tracked
File Edit Options Buffers Tools Python Help
import RPi.GPIO as GPIO
import time
from rrb2 import *

rr = RRB2()

def init_vehicle():
    rr.set_led1(1)

def turn_left():
    rr.set_motors(1, 1, 1, 0)

def turn_right():
    rr.set_motors(1, 0, 1, 1)

def forward():
    rr.set_motors(1, 1, 1, 1)

def backward():
    rr.set_motors(1, 0, 1, 0)

def stop():
    rr.set_motors(0, 0, 0, 0)

def cleanup():
    GPIO.cleanup()

-UU-:----F1   track.py        All L5      (Python)----------------
```

The following is the simple code to exercise this program using keystrokes on Raspberry Pi:

```
pi@raspberrypi: ~/robot                                    _  □  X
File Edit Options Buffers Tools Python Help
import RPi.GPIO as GPIO
import time
from rrb2 import *
from wheel import *
import tty
import sys
import termios
def getch():
    fd = sys.stdin.fileno()
    old_settings = termios.tcgetattr(fd)
    tty.setraw(sys.stdin.fileno())
    ch = sys.stdin.read(1)
    termios.tcsetattr(fd, termios.TCSADRAIN, old_settings)
    return ch
var = 'n'
while var != 'q':
    var = getch()
    if var == 'l':
        turn_right()
    if var == 'r':
        turn_left()
    if var == 'f':
        forward()
    if var == 'b':
        backward()
    if var == 's':
        stop()

GPIO.cleanup()

-UU-:----F1   input.py        All L1     (Python)------------------------
For information about GNU Emacs and the GNU system, type C-h C-a.
```

Run this program by typing `sudo python input.py`. If you get an error message telling you that the program does not know about `rr=RRB2()`, you'll need to copy the `rrb2.py` file from your installation of the libraries for the motor controller board in the directory `rrb2-1.1`.

Now that you can control your wheeled robot from the console, let's connect it wirelessly to allow remote control.

Controlling your robot from a remote computer

The next step in constructing your laser tag playing robot is to add the remote control. You'll first need to add the ability to control Raspberry Pi via a WLAN connection. As noted in *Chapter 1*, *Adding Raspberry Pi to an RC Vehicle*, the section, *Accessing the RC Car remotely*, showed you how to add a WLAN interface as an access point. To prepare for the next section, you will want to configure two Raspberry Pis with WLAN, one as an access point and other that you'll connect to the access point. You'll also connect a USB web camera to Raspberry Pi on the wheeled robot for control. Finally, when you have logged in to Raspberry Pi, this will function on the remote computer. Log in to the second Raspberry Pi on the wheeled robot by typing `ssh -X pi@xxx.xxx.xxx.xxx`, where the `xxx.xxx.xxx.xxx` is the IP address of Raspberry Pi on the wheeled robot.

You can now **luvcview** to see the output of the webcam and you can run the `input.py` program to control the wheeled robot. The output should look as follows:

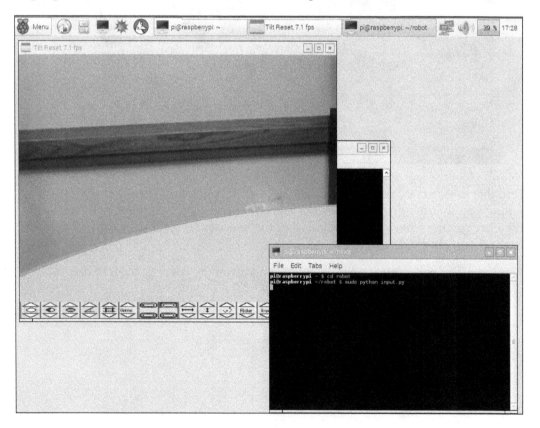

You can now type in `l`, `r`, `f`, `b` and control your robot. This is all well and good, but the frame rate is a bit slow. If you make the resolution smaller, the system will run faster. To do this type `luvcview -s 160x120` to adjust to the desired resolution. Now you can see and control your wheeled robot but you may want to add a game controller to your host computer to control your robot more intuitively.

Adding a game controller to your remote system

Typing the simple ascii characters will move your robot around, but what you might want is a more intuitive and responsive control interface. The most practical way of doing this is a game controller that has joysticks and several additional buttons. This will make controlling your wheeled robot from the remote computer much easier.

To add the game controller, you'll need to first find a game controller that can be connected to your computer. If you are using Microsoft Windows as the OS on the host computer, then pretty much any USB controller that can be connected to a PC, will work. The same type of controller also works if you are using Linux for the remote computer. In fact, in this example, you'll use another Raspberry Pi as the remote computer.

Since the joystick will be connected to the remote computer, you'll need to run two programs, one on Raspberry Pi remote computer and other on Raspberry Pi on the wheeled robot. You'll use the wireless LAN interface and a client-server model of communication. You'll run the server program on Raspberry Pi that is the remote computer and the client program on Raspberry Pi on the wheeled robot.

> For an excellent tutorial on this type of model and how it is used in a gaming application, refer to `http://www.raywenderlich.com/38732/multiplayer-game-programming-for-teens-with-python`.

Once you have the controller connected, you'll need to create a python program on Raspberry Pi that will receive the signals sent from the client and send the correct signals to the DC motors. This is the client program, but before you do that you'll need a LAN communication layer library called `PodSixNet`. This will allow the two applications to communicate. To install this, follow the instructions at `https://github.com/chr15m/PodSixNet/`. You'll need to install some tools that python will need to set up this capability by typing `sudo apt-get install python-setuptools` before installing this library, which is not in the documentation. Now you are ready to create the client program on Raspberry Pi on the wheeled platform. This will take the joystick commands that are sent from the server program running on the remote Raspberry Pi and translate them to commands that will control the wheeled robot.

The first part of the program shows the includes for the program and the main part of the `RobotGame` class that sends the commands to the DC motor controller. The laser fire print statements will be replaced later with a function call to fire the laser. The following is the first part of the code:

```
pi@raspberrypi: ~/robot

File Edit Options Buffers Tools Python Help
import pygame
import math
from PodSixNet.Connection import ConnectionListener, connection
from time import sleep
from wheel import *

class RobotGame(ConnectionListener):
    def Network_close(self, data):
        exit()
    def Network_gamepad(self, data):
        if data["type"] == 10:
            if data["info"]["button"] == 4:
                print "Fire Laser"
            if data["info"]["button"] == 5:
                print "Fire Laser"
            if data["info"]["button"] == 6:
                print "Fire Laser"
            if data["info"]["button"] == 7:
                print "Fire Laser"
        if data["type"] == 7:
            if data["info"]["value"] == 0.0:
                stop()
            else:
                if data["info"]["axis"] == 1:
                    if data["info"]["value"] > 0:
                        forward()
                    else:
                        backward()
                if data["info"]["axis"] == 2:
                    if data["info"]["value"] > 0:
                        turn_left()
                    else:
                        turn_right()
    def __init__(self):
        address=raw_input("Address of Server: ")
        try:
            if not address:
                host, port="localhost", 8000
            else:
                host,port=address.split(":")
            self.Connect((host, int(port)))
        except:
            print "Error Connecting to Server"
            print "Usage:", "host:port"
-UU-:**--F1  robot_client.py   Top L1      (Python)--------------
```

The `RobotGame` class does the actual command translation and sends the commands to the motor controller or the laser source (you'll learn how to hook up the laser source in the final section of this chapter.)

The following is a table of these controls:

Joystick control	Wheeled robot control
Button 4, 5, 6, or 7 (these are the trigger buttons on the front of the joystick)	Fire laser
Joystick 1 Up/Down	Forward/Backward
Joystick 1 Right/Left	Right/Left

Now, here is the second part of this class and the main body of the code:

```
                    else:
                        turn_right()
    def __init__(self):
        address=raw_input("Address of Server: ")
        try:
            if not address:
                host, port="localhost", 8000
            else:
                host,port=address.split(":")
            self.Connect((host, int(port)))
        except:
            print "Error Connecting to Server"
            print "Usage:", "host:port"
            print "e.g.", "localhost:31425"
            exit()
        print "Robot client started"
        self.running=False
        while not self.running:
            self.Pump()
            connection.Pump()
            sleep(0.01)

bg=RobotGame() #__init__ is called right here
while 1:
    if bg.update()==1:
        break
bg.finished()
```

`-UU-:**--F1  robot_client.py    Bot L46    (Python)--------------`

The second part of the code initializes the connection to the remote server. The last part of the code initializes the game loop, which loops while taking the input and sends it to the motor controller and the DC motors.

You'll also need a server program running on the remote computer that will take the signals from the game controller and send them to the client. You'll be writing this code in python using python version 2.7, which can be installed from here. Additionally, you'll also need to install the **pygame** library. If you are using Linux on the remote computer, then type `sudo apt-get install python-pygame`. If you are using Microsoft Windows on the remote machine, then follow the instructions given at `http://www.pygame.org/download.shtml`.

You'll also need the LAN communication layer described earlier. You can find a version that will run on Microsoft Windows or Linux at `https://github.com/chr15m/PodSixNet/`. The following is a listing of the server code in two parts:

```
Python 2.7.8: flightserver - C:/Python27/flightserver
File  Edit  Format  Run  Options  Windows  Help
import PodSixNet.Server
from pygame import *
from time import sleep
init()
from time import sleep
class ClientChannel(PodSixNet.Channel.Channel):
    def Network(self, data):
        print data
    def Close(self):
        self._server.close(self.gameid)
class BoxesServer(PodSixNet.Server.Server):

    channelClass = ClientChannel
    def __init__(self, *args, **kwargs):
        PodSixNet.Server.Server.__init__(self, *args, **kwargs)
        self.games = []
        self.queue = None
        self.currentIndex=0
    def Connected(self, channel, addr):
        print 'new connection:', channel
        if self.queue==None:
            self.currentIndex+=1
            channel.gameid=self.currentIndex
            self.queue=Game(channel, self.currentIndex)
    def close(self, gameid):
        try:
            game = [a for a in self.games if a.gameid==gameid][0]
            game.player0.Send({"action":"close"})
        except:
            pass
    def tick(self):
        if self.queue != None:
            sleep(.05)
            for e in event.get():
                self.queue.player0.Send({"action":"gamepad", "type":e.type, "in
        self.Pump()
class Game:
    def __init__(self, player0, currentIndex):
        #initialize the players including the one who started the game
        self.player0=player0

#Setup and init joystick
j=joystick.Joystick(0)
j.init()

#Check init status
if j.get init() == 1: print "Joystick is initialized"
```

This first part creates three classes, given as follows:

- The first class, `ClientChannel`, establishes a communication channel for your project
- The second class, `BoxesServer`, sets up a server so that you can communicate the joystick action to Raspberry Pi
- Finally, the `Game` class just initializes a game that contains everything that you'll need

The following is the second part of the code:

```
Python 2.7.8: flightserver - C:/Python27/flightserver
File  Edit  Format  Run  Options  Windows  Help
            sleep(.05)
            for e in event.get():
                self.queue.player0.Send({"action":"gamepad", "type":e.type, "in
        self.Pump()
class Game:
    def __init__(self, player0, currentIndex):
            #initialize the players including the one who started the game
            self.player0=player0

#Setup and init joystick
j=joystick.Joystick(0)
j.init()

#Check init status
if j.get_init() == 1: print "Joystick is initialized"

#Get and print joystick ID
print "Joystick ID: ", j.get_id()

#Get and print joystick name
print "Joystick Name: ", j.get_name()

#Get and print number of axes
print "No. of axes: ", j.get_numaxes()

#Get and print number of trackballs
print "No. of trackballs: ", j.get_numballs()

#Get and print number of buttons
print "No. of buttons: ", j.get_numbuttons()

#Get and print number of hat controls
print "No. of hat controls: ", j.get_numhats()


print "STARTING SERVER ON LOCALHOST"
# try:
address=raw_input("Host:Port (localhost:8000): ")
if not address:
    host, port="localhost", 8000
else:
    host,port=address.split(":")
boxesServe = BoxesServer(localaddr=(host, int(port)))

while True:
    boxesServe.tick()
    sleep(0.01)
```

This part of the code initializes the joystick so that all the controls can be sent to the wheeled robot's Raspberry Pi.

You'll need to run these programs on both computers by entering the IP address of the remote computer connected to the joystick. Here is what running the server program on the remote computer looks like:

```
pi@raspberrypi: ~
pi@raspberrypi ~ $ python joystick.py
Joystick is initialized
Joystick ID:  0
Joystick Name:   2603666 CONTROLLER
No. of axes:  4
No. of trackballs:  0
No. of buttons:   12
No. of hat controls: SDL_JoystickNumHats value:1:
 1
STARTING SERVER ON LOCALHOST
Host:Port (localhost:8000): 157.201.194.150:8000
```

And the following is how the program will look when it is run on Raspberry Pi connected to the wheeled robot:

```
pi@raspberrypi: ~/robot
pi@raspberrypi ~/robot $ sudo python robot_client.py
Address of Server: 157.201.194.150:8000
Robot client started
```

And finally, here is how the program will look on the remote computer when the robot's Raspberry Pi is up and connected:

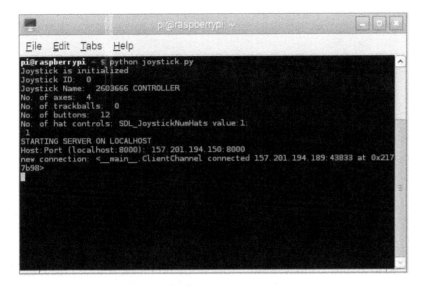

Now you can control your robot remotely using the joystick!

Connecting the laser source and target

The last step in creating the laser tag robots is to add the laser source and laser target, as well as the code that will let you fire the source and detect a hit on the target. First, you'll need to install the hardware.

Let's start with the laser source. You can just use a raw laser source, similar to the one shown in the following image:

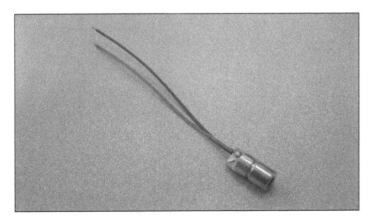

They are inexpensive and available on eBay and other online electronics retailers. However, it is a bit easier if you get a laser source with a bit more supporting circuitry, as shown in the following:

These are also available from a number of online retailers, you can find similar modules. All you need is a way to turn on your laser with a control signal from Raspberry Pi. In this module, you'll connect the device to one of the open collector outputs on the motor controller board, the **GND** pin to **OC1** pin, and the **S** pin to other **OC1** pin connection, as shown in the following image:

Now, just a bit of code to add a function to turn on the laser, as given in the following screenshot:

```
pi@raspberrypi: ~/robot
File Edit Options Buffers Tools Python Help
import RPi.GPIO as GPIO
import time
from rrb2 import *

rr = RRB2()

def laser_on():
    rr.set_led1(1)
    rr.set_oc1(1)
    time.sleep(1)
    rr.set_led1(0)
    rr.set_oc1(0)




-UU-:----F1  laser.py        All L1     (Python)-----------------------------
For information about GNU Emacs and the GNU system, type C-h C-a.
```

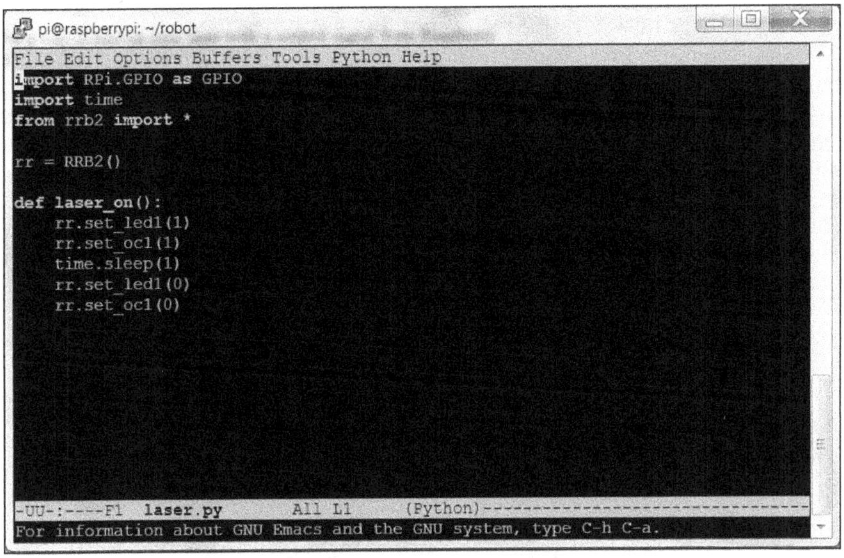

And finally, add the code we just saw to your main python file by importing it at the top of the file and then calling the function when a button is pressed, as given in the following screenshot:

```
pi@raspberrypi: ~/robot
File Edit Options Buffers Tools Python Help
import pygame
import math
from PodSixNet.Connection import ConnectionListener, connection
from time import sleep
from wheel import *
from laser import *

class RobotGame(ConnectionListener):
    def Network_close(self, data):
        exit()
    def Network_gamepad(self, data):
        if data["type"] == 10:
            if data["info"]["button"] == 4:
                laser_on()
                print "Fire Laser"
            if data["info"]["button"] == 5:
                laser_on()
                print "Fire Laser"
            if data["info"]["button"] == 6:
                laser_on()
                print "Fire Laser"
            if data["info"]["button"] == 7:
                laser_on()
                print "Fire Laser"
        if data["type"] == 7:
            if data["info"]["value"] == 0.0:
                stop()
            else:
-UU-:----F1  robot_client.py   Top L1    (Python)-----------------------------
For information about GNU Emacs and the GNU system, type C-h C-a.
```

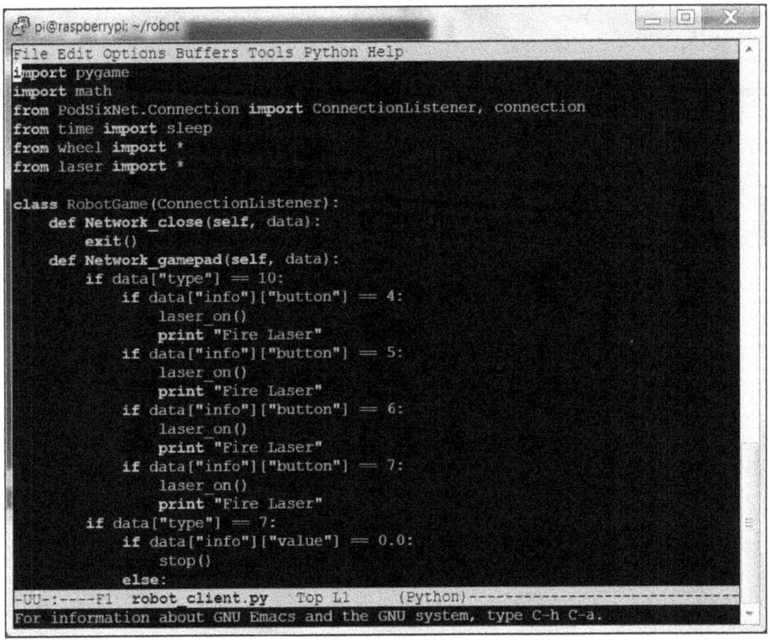

Now when a button is pressed on the joystick, the laser should turn on. The final step in enabling your laser tag robot is to add the target. In this case, to keep it simple, you'll add a target device and simple python program, which you can run in another window, that will signal when the target has been hit.

There are several possible approaches to add a target. You can build your own target array using photo-sensitive resistors, as shown in the following image:

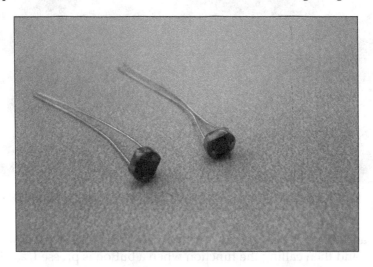

This is a bit difficult as Raspberry Pi doesn't have an **analog-to-digital convertor** (**ADC**). An easier way to do this is to add a simple laser sensor, similar to the one shown in the following image:

This target will sense when the laser has struck and then output a signal so that you know that you have been tagged. This particular device also has a laser source that is always on, but you'll want to put a piece of electrical tape over the source or it will give you away as you try to outwit your opponent. In this format, the sensor almost has a 180 degree sensitivity but you may want to put two sensors back to back if you want to be able to target the front and back of your wheeled robot simultaneously.

This device has three connections, a **VCC**, a **GND**, and a **DOUT**. You'll connect the **VCC** connector to the **Pin 1 3.3V** connector, the **GND** connector to the **Pin 6 GND** connector, and the **DOUT** connector to the **Pin 18 GPIO 24** connector of the **general purpose input/output** (**GPIO**) pins of Raspberry Pi. The following is a layout of these pins:

Pin 1 3.3V	▢ ◯	Pin 2 5V
Pin 3 GPIO2	◯ ◯	Pin 4 5V
Pin 5 GPIO3	◯ ◯	Pin 6 GND
Pin 7 GPIO4	◯ ◯	Pin 8 GPIO14
Pin 9 GND	◯ ◯	Pin 10 GPIO15
Pin 11 GPIO17	◯ ◯	Pin 12 GPIO18
Pin 13 GPIO27	◯ ◯	Pin 14 GND
Pin 15 GPIO22	◯ ◯	Pin 16 GPIO23
Pin 17 3.3V	◯ ◯	Pin 18 GPIO24
Pin 19 GPIO10	◯ ◯	Pin 20 GND
Pin 21 GPIO9	◯ ◯	Pin 22 GPIO25
Pin 23 GPIO11	◯ ◯	Pin 24 GPIO8
Pin 25 GND	◯ ◯	Pin 26 GPIO7
Pin 27 ID_SD	◯ ◯	Pin 28 ID_SC
Pin 29 GPIO5	◯ ◯	Pin 30 GND
Pin 31 GPIO6	◯ ◯	Pin 32 GPIO12
Pin 33 GPIO13	◯ ◯	Pin 34 GND
Pin 35 GPIO19	◯ ◯	Pin 36 GPIO16
Pin 37 GPIO26	◯ ◯	Pin 38 GPIO20
Pin 39 GND	◯ ◯	Pin 40 GPIO21

The connections will look as shown in the following image:

Now for the code. It is quite simple, you'll query the input and when it goes from 1 to 0, you have a signal indicating that the laser has connected with your target. You register it, as shown in the following screenshot:

```python
import RPi.GPIO as GPIO
import time

GPIO.setmode(GPIO.BCM)
target_pin = 24
GPIO.setup(target_pin,GPIO.IN)
while 1:
    hit = GPIO.input(target_pin)
    if hit == 0:
        print("HIT detected");
    time.sleep(.1)
```

Now you'll mount your laser on the wheeled robot and laser target at the same height for all the combatants; access your wheeled robot remotely; open the webcam window, joystick control windows, and target detected window; and you're ready for a good game of laser tag.

Summary

This chapter's result is a laser tag playing machine that can be guided remotely. You can even build a simple battlefield with different maze elements and let your wheeled robots loose to play a game of laser tag. In the next chapter, you'll move onto something quite different, a machine that can draw.

5
A Robot That Can Draw

You've modified several toys, built a wheeled and tracked platform, and made them all do amazing things. In this chapter, you'll move from a mobile platform to a fixed one, with a specific goal in mind; to build a robot that can draw.

In this chapter, you'll learn:

- Constructing a drawing platform using servos and brackets
- Using a servo controller to control multiple servos
- Creating a Python program to control servos
- Using servos to create a drawing robot
- Connecting to a graphical Python program to control the movement of the drawing robot

Constructing a drawing platform using servos and brackets

To begin the project, you'll first need to build a robot arm to do the drawing. There are several robotic arms, which are available at many online electronics outlets and eBay, that would do well for this application. A less expensive approach would be to use a set of servo brackets and construct your robotic arm. The following is an image of this arm:

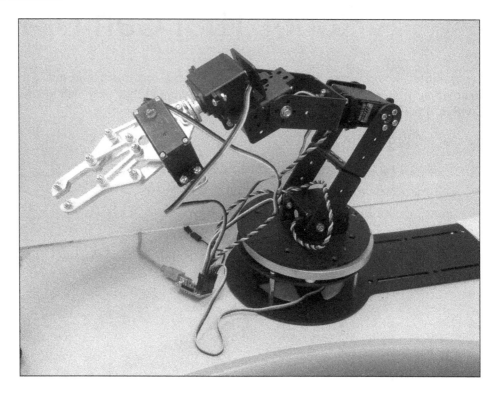

To construct this arm, you can purchase a set of servo brackets on eBay or look for a kit that uses a six **degrees of freedom** (**DOF**) mechanical robotic arm at any one of the several online electronics retailers. You can either build the kit or an arm of your own design. Remember to make sure that you get servos with enough torque. For the upper end of the arm, almost any servo will do. For the servo attached at the base and up the chain, you'll need servos with fairly significant torque capabilities. I like to use the **HS-645MG** servos by **Hitec** for this application, they have metal internal gearing and a torque rating of 133 oz/in when used with a 6V source.

There are other configurations that are excellent for drawing. Refer to http://www.instructables.com/id/Drawing-Robot/ or http://blog.makeblock.cc/makeblock-drawing-robot/ for examples. However, the robot arm that you are constructing is more of a general purpose arm and can be repurposed for other tasks as well.

For the robot arm that you are building, you will need to control six different servos. You could control a single servo using Raspberry Pi directly but since you're going to control six of them, you'll want to use an external servo controller. Here is an image of a six servo controller made by **Pololu**, available at https://www.pololu.com/ and other online electronics retailers:

To make your robotic arm move, you first need to connect the servo motor controller to the servos. There are two connections that you need to make. The first is to the servo motors and the second is to the battery. In this section, before connecting your controller to your Raspberry Pi, you'll need connect your servo controller to your PC or Linux machine to check whether everything is working or not.

But first, you'll need to connect the servos to the controller.

Connect your six servos to the connections that are marked from 0 to 5 on the controller using the following configurations:

- **0**: The servo to turn the base
- **1**: The servo to control the up/down motion of the entire arm
- **2**: The servo at the elbow of the arm
- **3**: The servo to move the wrist up and down
- **4**: The servo to turn the wrist
- **5**: The servo to open and close the claw

The following is an image of the back of the controller; it will show you where to connect your servos:

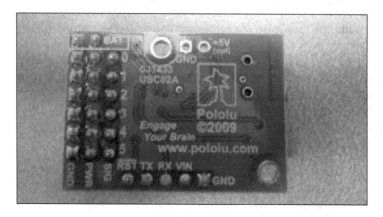

The following image illustrates the servos that are connected to the controller:

Here are the steps to connect the board to the power supply:

1. Now you need to connect the servo motor controller to your power supply. If you have a USB style power connection, you can use FTDI's USB to UART cable and then plug the red and black cables into the power connector on the servo controller, as shown in the following image:

2. Now, plug the other end of the USB cable into the USB port of the power connection, as follows:

The hardware is now ready to control!

Configuring the software

Now, you can connect the motor controller to your PC or Linux machine to see whether or not you can talk to it. Once the hardware is connected, you will use some of the software provided by Pololu to control the servos. Download the Pololu software from `http://www.pololu.com/docs/0J40/3.a` and install it using the instructions given on the website. Once it is installed, run the software; you should be able to see the window that is shown in the following screenshot:

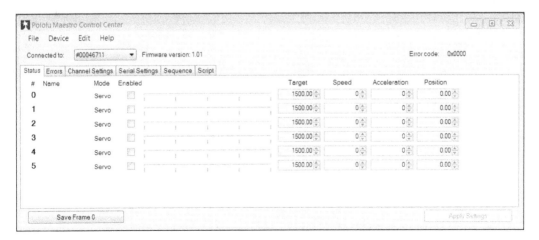

You will first need to change the **Serial mode** configuration in **Serial Settings**, so select the **Serial Settings** tab; you will see the window that is shown in the following screenshot:

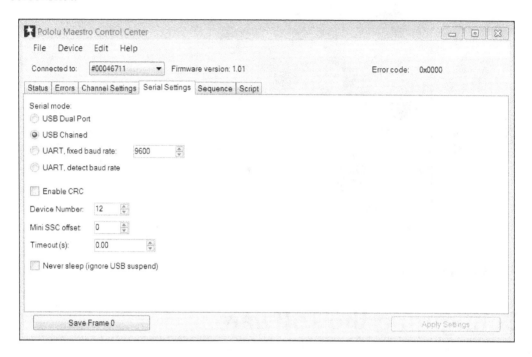

Make sure that **USB Chained** is selected; this will allow you to connect to and control the motor controller over the USB. Now, go back to the main screen by selecting the **Status** tab and you can now turn on the four servos. The screen will look as shown in the following screenshot:

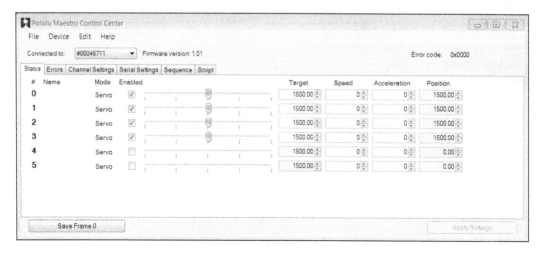

Now, you can use the sliders to control the servos. Enable the four servos and make sure that servo **0** moves the base; **1**, the lowest servo; **2**, the elbow of the arm; **3**, the up and down motion of the wrist, **4**, the roll of the wrist, and **5**, open and close the claw.

You've checked the motor controllers and servos, now you'll connect the motor controller to Raspberry Pi to control the servos from there. Remove the USB cable from the PC and connect it to Raspberry Pi. The entire system will look similar to the following image:

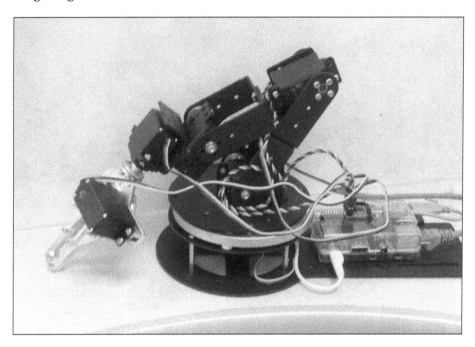

Let's now talk to the motor controller from your Raspberry Pi by downloading the Linux code from Pololu at `http://www.pololu.com/docs/0J40/3.b`. To do this, follow these steps:

1. First log on to Raspberry Pi using PuTTY and download the `maestro_linux_150116.tar.gz` file. To do this, type `wget http://www.pololu.com/file/download/maestro-linux-150116.tar.gz?file_id=0J315` into a terminal window.

2. To move this file into a file that can be used, type `mv maestro-linux-150116.tar.gz\?file_id\=0J315 maestro-linux-150118.tar.gz`. Unpack the file by typing `tar -xzfv maestro_linux_150116.tar.gz`. This will create a folder called `maestro_linux`. Go to this folder by typing `cd maestro_linux` and then type `ls`. You will see the output as shown in the following screenshot:

```
pi@raspberrypi: ~/maestro_linux
pi@raspberrypi ~/maestro_linux $ ls
99-pololu.rules  FirmwareUpgrade.dll  README.txt    UsbWrapper.dll  Usc.dll
Bytecode.dll     MaestroControlCenter  Sequencer.dll  UscCmd
pi@raspberrypi ~/maestro_linux $
```

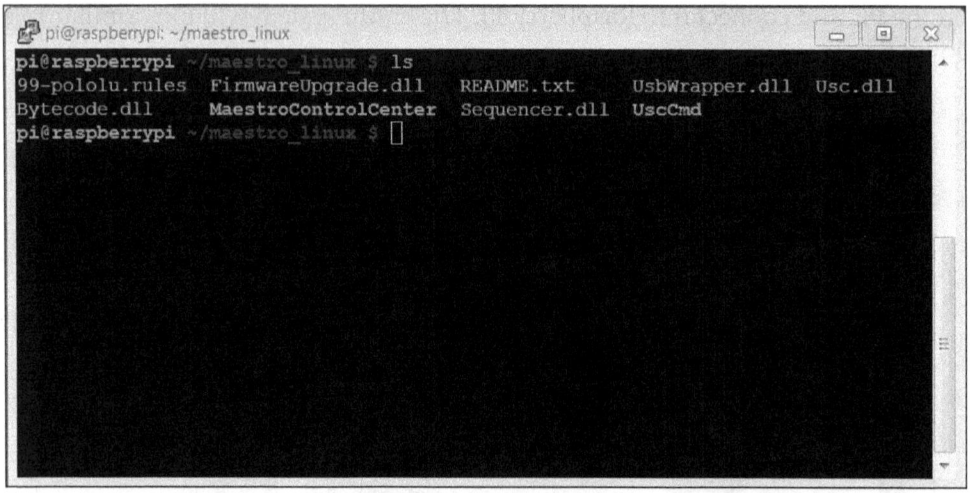

3. The `README.txt` document will give you explicit instructions on how to install the software. Unfortunately, you can't run **Maestro Control Center** on your Raspberry Pi. The version of Maestro Control Center that is considered doesn't support the Raspberry Pi graphical system, but you can control your servos using the **UscCmd** command-line application. First, type `./UscCmd --list`; you will see the following screenshot:

```
pi@raspberrypi: ~/maestro_linux
pi@raspberrypi ~/maestro_linux $ ./UscCmd --list
1 Maestro USB servo controller device found:
#00027392
pi@raspberrypi ~/maestro_linux $
```

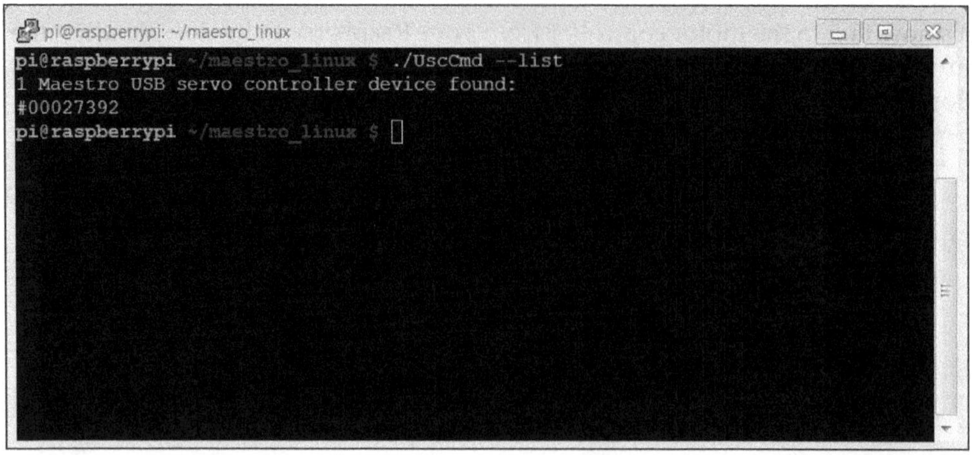

The software now recognizes that you have a servo controller. If you just type
`./UscCmd`, you can see all the commands that you could send to your controller.
When you run this command, you can see the result that is shown in the following
screenshot:

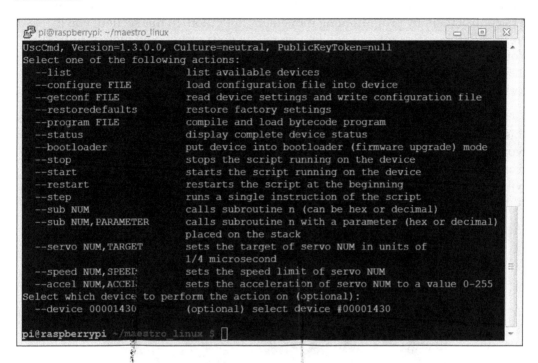

Notice that you can send a specific target angle to a servo, although if the target
angle is not within the range, this makes it a bit difficult to know where you are
sending your servo. Try typing `./UscCmd --servo 0, 10`. The servo will most
likely move to its full angle position. Type `./UscCmd --servo 0, 0` and it will stop
the servo from trying to move. In the next section, you'll write software that will
translate your angles into the electronic signals that will move the servos.

If you haven't run the Maestro Controller tool and set the **Serial Settings** setting to
USB Chained, your motor controller may not respond.

Creating a program in Python to control the mobile platform

Now that you can control your servos by using a basic command-line program, let's control them by programming movement in Python. In this section, you'll create a Python program that will let you talk to your servos a bit more intuitively. You'll issue commands that will tell a servo to go to a specific angle and it will go to that angle. You can then add a set of such commands to allow your robot to move forward, backward, or position the claw to any specific location.

Let's start with a simple program that will make your robot's servos turn 90 degrees; this will be somewhere close to the middle of the 180 degree range that you can work within. However, the center, maximum, and minimum values can vary from one servo to another, so you may need to calibrate them. To keep things simple, we will not cover this here. The following screenshot shows the code that is required for turning the servos:

```
pi@raspberrypi: ~/maestro-linux
File Edit Options Buffers Tools Python Help
#!/usr/bin/python
import serial
import time
def setAngle(ser, channel, angle):
    minAngle = 0.0
    maxAngle = 180.0
    minTarget = 256.0
    maxTarget = 13120.0
    scaledValue = int((angle / ((maxAngle - minAngle) / (maxTarget - minTarget)\
)) + minTarget)
    commandByte = chr(0x84)
    channelByte = chr(channel)
    lowTargetByte = chr(scaledValue & 0x7F)
    highTargetByte = chr((scaledValue >> 7) & 0x7F)
    command = commandByte + channelByte + lowTargetByte + highTargetByte
    ser.write(command)
    ser.flush()
def setSpeed(ser, channel, speed):
    if speed > 127 or speed <0:
        speed=1
    commandByte = chr(0x87)
    channelByte = chr(channel)
    highByte, lowByte = divmod(speed, 32)
    highTargetByte = chr(highByte)
    lowTargetByte = chr(lowByte << 2)
    command = commandByte + channelByte + lowTargetByte + highTargetByte
    ser.write(command)
    ser.flush()
def setHome(ser):
    for i in range(0, 5):
        setAngle(ser, i ,90)

ser = serial.Serial("/dev/ttyACM0", 9600)
setHome(ser)
time.sleep(1)
while 1:
    servo = int(raw_input("Servo number: "))
    angle = int(raw_input("Angle: "))
    speed = int(raw_input("Speed: "))
    setSpeed(ser, servo, speed)
    setAngle(ser, servo, angle)
    time.sleep(.5)
-UU-:----F1  robotArm.py    All L1      (Python)-----------------------
For information about GNU Emacs and the GNU system, type C-h C-a.
```

The following is an explanation of the code:

- The `#!/user/bin/python` line allows you to make this Python file available for execution from the command line. It will also allow you to call this program from your voice command program. We'll talk about this in the next section.

- The `import serial` and `import time` lines include the serial and time libraries. You need the serial library to talk to your unit via USB. If you have not installed this library, type `sudo apt-get install python-serial`. You will use the `time` library later to wait between the servo commands.

- The `setAngle()` function converts your desired settings for the servo and angle to the serial command that the servo motor controller requires. The values — `minTarget` and `maxTarget` — and the structure of the communications — `channelByte`, `commandByte`, `lowTargetByte`, and `highTargetByte` — come from the manufacturer.

- The `setSpeed()` method sets the speed of the movement. This function converts your desired settings for the specific servo and speed for that servo to the serial command that the servo motor controller requires. The values, such as the structure of the communications — `channelByte`, `commandByte`, `lowTargetByte`, and `highTargetByte` — come from the manufacturer.

- The `setHome` function moves all the servos to the 90 degree location.

- The last part of the code sets up the serial ports, sets all the servos to the home location, allows the user to enter a servo position and speed, and then sets the servo to that position at that speed.

Now you can set each servo to the desired position at the desired speed. The default would be to set each servo to a 90 degree angle. However, the servos were exactly centered, so you may realize that you need to move the servo horn where you want the servos to be centered.

Once you have the basic home position set, you can now ask your robot arm to do some things, such as, to begin drawing.

Simple drawing using the robotic arm

You can now ask your arm to hold a drawing pen by simply opening the claw, inserting a pen, and closing the claw. You'll probably want to use some sort of marker or other drawing device with a wide tip and lots of color. The following is the arm holding the pen:

If you adjust the servo that moves the entire arm up and down, you can make the pen touch the paper, as follows:

Finally, if you move the servo that turns the base, you can draw your first curved line, similar to the following image:

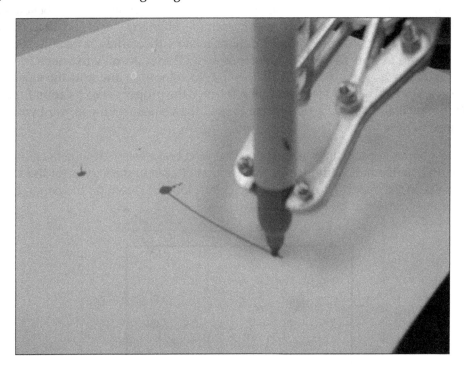

This is wonderful, but drawing only curved lines will not be particularly useful; besides, an interface where the user enters the servo locations one at a time to try and draw a more complex drawing will be unacceptable. What you need is a program that takes an x-y location input and moves the robotic arm to that point with the proper servo positions.

Since the input coming from the computer will be the x and y location for a point, you'll need to translate that value into the proper angle positions for your arm. This can be done in two ways; first, you can build a translation lookup table for the servo locations for each x and y location on the paper and second, you can model the positions of the servo using mathematical formulas to determine the servo locations.

If you want to build a translation table, one key question is how many individual *x* and *y* locations you'll want to define for your robotic arm. With more points, the table will be larger, but you'll also be able to draw a finer resolution. It is best to start with a table that isn't particularly fine, for example, 60 in the y axis and 40 in the x axis. This gives you 2,400 individual locations to store the values of the four servos. You'll not need to store the values for servos 4 and 5, they won't vary much for each implementation. You can simply walk through each location and note the value of the servos, store these in a table, and then retrieve the proper servo location for each position. You'll also want a similar table to hold the values when you want your robot arm to raise the pen from the paper.

Perhaps a better approach, or at least one that won't take as much time to calibrate, is to build a mathematical model for the position of each of the servos for the *x* and *y* locations on the paper. The following is a view, looking down on the paper, of the four corners as marked by the robotic arm:

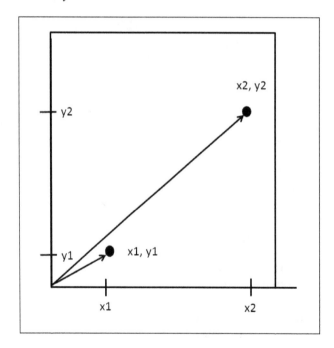

If you think of the arm being placed in the lower left corner, you can use mathematics to build a model of how to turn the angle of the position of the servo at the base of the arm to position it for two different points, as shown in the following diagram:

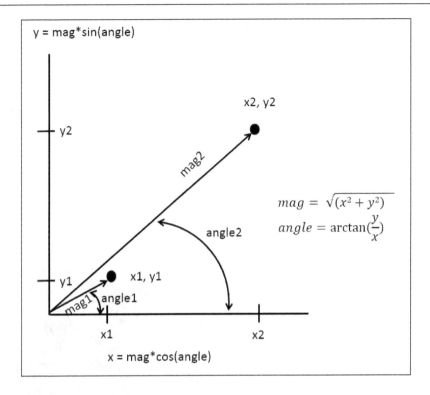

These equations show the relationship among the point x, point y, angle, and distance to the point. The angle calculation here can be used to position the servo at the base of the arm, servo 0. Adjusting the magnitude is a bit more difficult, you'll be using servo 2 to adjust the distance of the arm to the point. This calculation will look something similar to the following diagram:

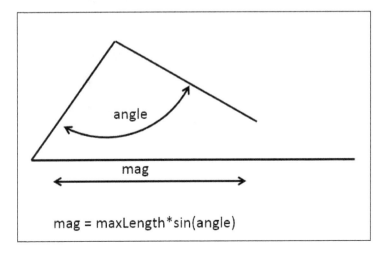

In this case, **maxLength** is the length of the arm when the angle of servo 2, at the elbow joint of the arm, is 90 degrees. Now there is actually the opportunity to go further but making this the max length will keep your calculations simple. The final calculation will be the calculation of the angle of servo 1, which can then raise the entire arm up and down. This servo will need to be set such that, based on the length of the arm, the servo will need to be raised or lowered to set this value. This value will also be set using a cos/sin equation, but based on the cos/sin of the angle of servo 2.

To code this, start by putting the setAngle() and setSpeed() functions into a library program, as shown in the following screenshot:

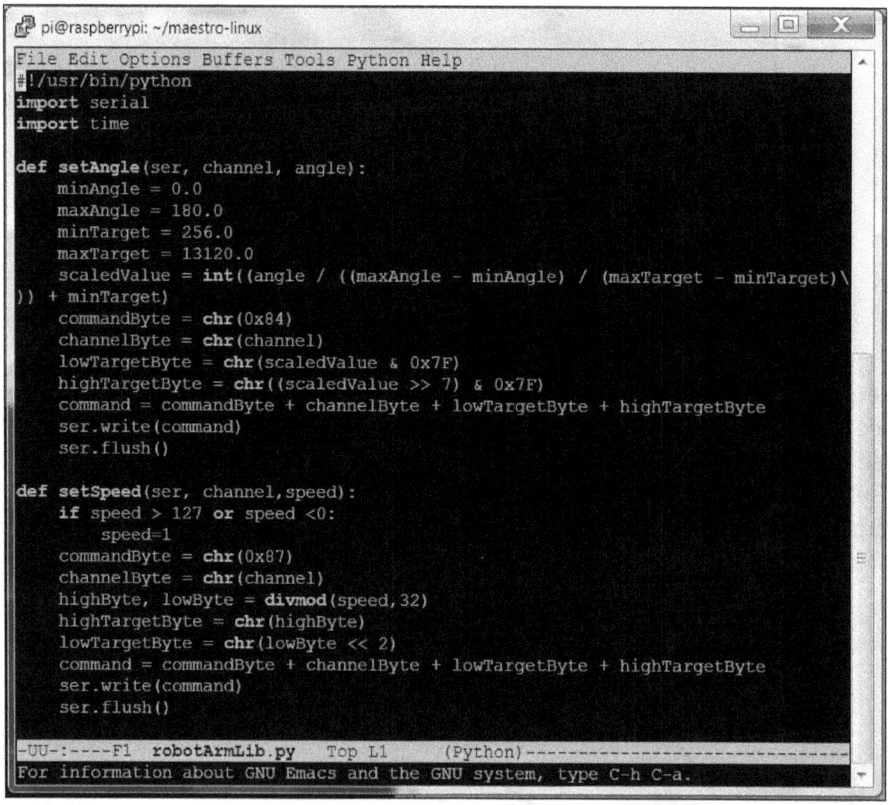

```python
#!/usr/bin/python
import serial
import time

def setAngle(ser, channel, angle):
    minAngle = 0.0
    maxAngle = 180.0
    minTarget = 256.0
    maxTarget = 13120.0
    scaledValue = int((angle / ((maxAngle - minAngle) / (maxTarget - minTarget)\
)) + minTarget)
    commandByte = chr(0x84)
    channelByte = chr(channel)
    lowTargetByte = chr(scaledValue & 0x7F)
    highTargetByte = chr((scaledValue >> 7) & 0x7F)
    command = commandByte + channelByte + lowTargetByte + highTargetByte
    ser.write(command)
    ser.flush()

def setSpeed(ser, channel, speed):
    if speed > 127 or speed < 0:
        speed=1
    commandByte = chr(0x87)
    channelByte = chr(channel)
    highByte, lowByte = divmod(speed, 32)
    highTargetByte = chr(highByte)
    lowTargetByte = chr(lowByte << 2)
    command = commandByte + channelByte + lowTargetByte + highTargetByte
    ser.write(command)
    ser.flush()
```

-UU-:----F1 **robotArmLib.py**   Top L1   (Python) ---------------------
For information about GNU Emacs and the GNU system, type C-h C-a.

Now, you'll create a simple program that can position the base servo using the *x* and *y* locations, as shown in the following screenshot:

```
pi@raspberrypi: ~/maestro-linux
File Edit Options Buffers Tools Python Help
#!/usr/bin/python
import serial
import time
import math
from robotArmLib import *

ser = serial.Serial("/dev/ttyACM0", 9600)

while 1:
    x = int(raw_input("x: "))
    y = int(raw_input("y: "))
    speed = 5
    servo0 = 0
    setSpeed(ser, servo0, speed)
    angle = int((math.atan2(y, x)* 360)/(2 * 3.1416))
    print angle
    setAngle(ser, servo0, angle + 40)
    time.sleep(.5)

-UU-:**--F1   robotPos.py    All L18    (Python)----------------------------
```

You'll use the `robotArmLib` library and its `setSpeed()` and `setAngle()` functions to set the position of the arm. Notice that this is all based on the x = 0 and y = 0 locations being at the center of the arm. On entering x = 50 and y = 0, the arm will position itself as shown in the following image:

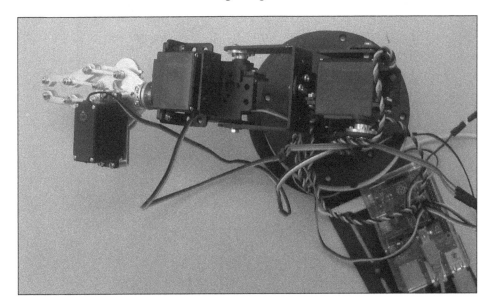

If you enter x = 0 and y = 50, you will see the arm in the following position:

And finally, setting x = 50 and y = 50 will move the arm to the following position:

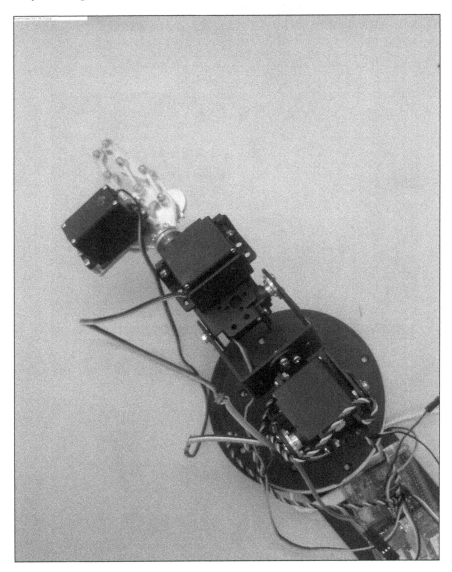

Now that you can position the base angle correctly, you'll need to position servo 2, at the elbow joint, to move in and out based on the distance from the center of the base. This code adds the control by adjusting servo 2 based on the magnitude of the distance:

```python
#!/usr/bin/python
import serial
import time
import math
from robotArmLib import *

ser = serial.Serial("/dev/ttyACM0", 9600)
while 1:
    x = int(raw_input("x: "))
    y = int(raw_input("y: "))
    speed = 5
    servo0 = 0
    servo2 = 2
    setSpeed(ser, servo0, speed)
    setSpeed(ser, servo2, speed)
    angle = int((math.atan2(y, x)* 360)/(2 * 3.1416))
    mag = int(math.hypot(x, y))
    print angle
    print mag
    setAngle(ser, servo0, angle + 40)
    setAngle(ser, servo2, mag + 40)
    time.sleep(.5)
```

So, for x = 10 and y = 0, the arm would be positioned as shown in the following image:

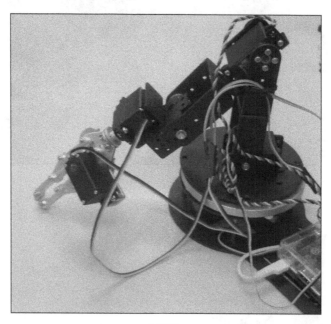

And, for x = 100 and y = 0, the arm would be positioned as follows:

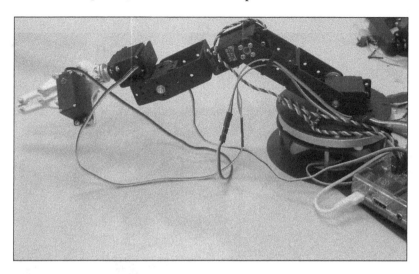

Now, you will need to change the servo setting for servo 1, at the base of the arm, to move the position so that the pen is on the paper. This angle will be controlled in a similar fashion to servo 2. Here is the entire code:

```python
#!/usr/bin/python
import serial
import time
import math
from robotArmLib import *

ser = serial.Serial("/dev/ttyACM0", 9600)

while 1:
    x = int(raw_input("x: "))
    y = int(raw_input("y: "))
    speed = 5
    servo0 = 0
    servo1 = 1
    servo2 = 2
    setSpeed(ser, servo0, speed)
    setSpeed(ser, servo2, speed)
    angle = int((math.atan2(y, x) * 360) / (2 * 3.1416))
    mag = int(math.hypot(x, y))
    print angle
    mag2 = mag/2
    print mag2
    setAngle(ser, servo0, angle + 40)
    setAngle(ser, servo2, 160 - mag)
    setAngle(ser, servo1, 50 + mag2)
    time.sleep(.5)
```

Using this code and asking for x = 10 and y = 0 now gives the total arm position as follows:

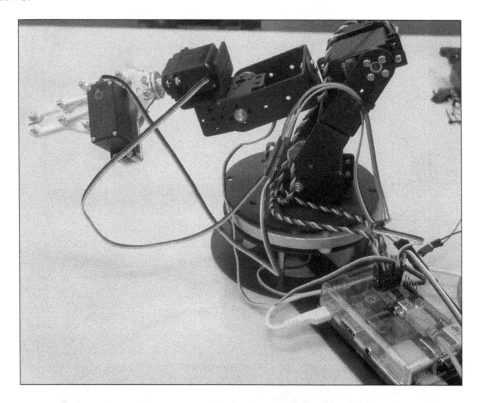

And, for x = 100 and y = 0, the total arm position is shown in the following image:

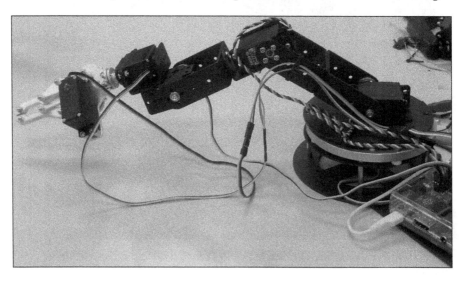

Now you can position the pen for each individual *x* and *y* location. You will, however, notice that the entire space between 100 and 10 is not linear and, as you get closer to the arm, it is more difficult to keep the pen on the paper. You should evaluate the available drawing distance, it may be between 100 and 50. The best way to evaluate this is to place a pen in the robot's hand and simply begin to draw the points in order to understand the boundaries of the robot's workable range.

This final Python code adds a loading function in order to load the pen and position it in the robot's claw:

```
#!/usr/bin/python
import serial
import time
import math
from robotArmLib import *

def loadPen(ser):
    servo = 1
    print "Loading Pen"
    print "Enter servo 0 when loaded"
    print "Servo 5 - Claw"
    print "Servo 4 - Wrist"
    while servo != 0:
        servo = int(raw_input("Servo number: "))
        angle = int(raw_input("Angle: "))
        speed = 5
        setSpeed(ser, servo, speed)
        setAngle(ser, servo, angle)
        time.sleep(.1)

ser = serial.Serial("/dev/ttyACM0", 9600)
loadPen(ser)
while 1:
    x = int(raw_input("x: "))
    y = int(raw_input("y: "))
    speed = 5
    servo0 = 0
    servo1 = 1
    servo2 = 2
    setSpeed(ser, servo0, speed)
    setSpeed(ser, servo2, speed)
    angle = int((math.atan2(y, x)* 360)/(2 * 3.1416))
    mag = int(math.hypot(x, y))
    print angle
    mag2 = mag/2
    print mag2
    setAngle(ser, servo0, angle + 40)
    setAngle(ser, servo2, 160 - mag)
    setAngle(ser, servo1, 50 + mag2)
    time.sleep(.5)
```

You may want to tape the pen in place, this can solidify its connection to
the arm.

The next step is to add an application that will control the program more elegantly
than simply typing in the *x* and *y* locations.

A simple Python drawing program

Now that the robot can draw, you can add a simple graphical program that allows
you to draw on the screen and then output this set of points to the drawing robot.
Let's start with a simple draw program that is based on **pygame**:

```
pi@raspberrypi: ~/maestro-linux
File Edit Options Buffers Tools Python Help
import pygame, sys

black = 0,0,0
white = 0xFF, 0xFF, 0xFF
background_color = 0x12, 0x0E, 0x1C

width = 64
height = 32
scale = 8
canvas = pygame.Surface((width, height), pygame.SRCALPHA)
canvas.set_at((0,0), white)
canvas.set_at((width-1,height-1), black)

def animation_frame(screen):
    screen.fill(background_color)
    view = pygame.transform.scale(canvas, (width*scale, height*scale))
    screen.blit(view, (0, 0))

def plot((x,y)):
    x = int(x/scale)
    y = int(y/scale)
    print x, y
    if 0 <= x < width and 0 <= y < height:
        canvas.set_at((x,y), white)

def dispatch(event):
    if event.type == pygame.QUIT:
        sys.exit(0)
    if event.type == pygame.MOUSEBUTTONDOWN:
        plot(event.pos)
    if event.type == pygame.MOUSEMOTION and event.buttons != (0,0,0):
        plot(event.pos)

pygame.display.init()
screen = pygame.display.set_mode((320, 240))
while 1:
    for event in pygame.event.get():
        dispatch(event)
    animation_frame(screen)
    pygame.display.flip()

-UU-:----F1   draw.py        All L1      (Python)----------------------------
For information about GNU Emacs and the GNU system, type C-h C-a.
```

When you run this program, either directly with a monitor and keyboard connected to Raspberry Pi or with the VNC Server viewer, you will see the following:

As you move the mouse in the draw window, you will notice the *x* and *y* location print in the terminal window. What you'll now do is save the robot arm control code in a library that can be called directly from the draw program. Here is the library code:

```python
#!/usr/bin/python
import serial
import time
import math
from robotArmLib import *

def loadPen(ser):
    servo = 1
    print "Loading Pen"
    print "Enter servo 0 when loaded"
    print "Servo 5 - Claw"
    print "Servo 4 - Wrist"
    while servo != 0:
        servo = int(raw_input("Servo number: "))
        angle = int(raw_input("Angle: "))
        speed = 5
        setSpeed(ser, servo, speed)
        setAngle(ser, servo, angle)
        time.sleep(.1)

def setPos(ser, x, y):
    speed = 5
    servo0 = 0
    servo1 = 1
    servo2 = 2
    setSpeed(ser, servo0, speed)
    setSpeed(ser, servo2, speed)
    angle = int((math.atan2(y, x)* 360)/(2 * 3.1416))
    mag = int(math.hypot(x, y))
    print angle
    mag2 = mag/2
    print mag2
    setAngle(ser, servo0, angle + 40)
    setAngle(ser, servo2, 160 - mag)
    setAngle(ser, servo1, 50 + mag2)
```

-UU-:----F1 **robotPosLib.py**   All L25   (Python)------------------------------
Wrote /home/pi/maestro-linux/robotPosLib.py

The following is the new draw code with the library connected:

```
pi@raspberrypi: ~/maestro-linux
File Edit Options Buffers Tools Python Help
import pygame, sys
from robotPosLib import *

black = 0,0,0
white = 0xFF, 0xFF, 0xFF
background_color = 0x12, 0x0E, 0x1C
ser = serial.Serial("/dev/ttyACM0", 9600)

width = 64
height = 32
scale = 8
canvas = pygame.Surface((width, height), pygame.SRCALPHA)
canvas.set_at((0,0), white)
canvas.set_at((width-1,height-1), black)

def animation_frame(screen):
    screen.fill(background_color)
    view = pygame.transform.scale(canvas, (width*scale, height*scale))
    screen.blit(view, (0, 0))

def plot((x,y)):
    x = int(x/scale)
    y = int(y/scale)
    print x, y
    setPos(ser, x + 50, y)
    if 0 <= x < width and 0 <= y < height:
        canvas.set_at((x,y), white)

def dispatch(event):
    if event.type == pygame.QUIT:
        sys.exit(0)
    if event.type == pygame.MOUSEBUTTONDOWN:
        plot(event.pos)
    if event.type == pygame.MOUSEMOTION and event.buttons != (0,0,0):
        plot(event.pos)

pygame.display.init()
screen = pygame.display.set_mode((320, 240))
while 1:
    for event in pygame.event.get():
        dispatch(event)
    animation_frame(screen)
    pygame.display.flip()

-UU-:----F1  robotDraw.py    All L1      (Python)--------------------------
```

Now, you can draw on the canvas and your robot arm will follow that set of motions. Of course, drawing is just one activity that your robot arm can tackle, there are myriad other activities that can utilize your robot arm.

Summary

You now know how to control the servos and move a robotic arm! In the next chapter, you'll take on a different kind of positioning system—the stepper motor—to build a robot that can play air hockey.

6
A Robot That Can Play Air Hockey

By now, you should have some amazing projects on your shelf, including projects that can walk, talk, and draw. Now, let's build a robot that can play air hockey.

In this chapter, you'll learn the following:

- How to build an air hockey robot paddle using three-dimensional printing concepts of gears and pulleys
- How to use Raspberry Pi and Arduino with the stepper motor drivers to control stepper motors
- How to connect a USB webcam and **OpenCV** to track the color and movement
- How to tie all this together for an unbeatable air hockey opponent

Constructing the platform

Constructing the hardware and connecting it to the table is a significant challenge. Fortunately, there is an excellent website that explains how to construct the entire hardware system using three-dimensional printed parts at `http://jjrobots.com/air-hockey-robot-a-3d-printer-hack/`. This website even provides you with the opportunity to purchase the three-dimensional printed parts. Follow the detailed instructions to build the hardware. Here is a picture of the hardware that is connected to a small air hockey table that was purchased at a local toy store:

In the documented example, the controller is a laptop that is connected to a **PlayStation 3 (PS3)** Eye camera. For this project, you'll replace the laptop and PS3 camera with Raspberry Pi and a webcam. And instead of having Arduino calculate the paddle position, you'll be using Raspberry Pi. To understand how to do this, you'll first need to understand how to control stepper motors with Arduino and stepper motor drivers.

Controlling the paddle using stepper motors

The first step in controlling your air hockey playing robot is to control the position of the three stepper motors in your paddle system. To understand how to do this, let's start by exploring how stepper motors work. The following is an image of a stepper motor:

Stepper motors are a bit different from the servos that you used in *Chapter 5, A Robot That Can Draw*. Stepper motors operate in a similar way to DC motors as even they can rotate continuously. However, the stepper motor has the ability to drive the motor in small steps. For a tutorial on the specifics on stepper motors, refer to https://en.wikipedia.org/wiki/Stepper_motor.

Stepper motors can be a bit difficult to work with as the control is more complex than DC or servo motors. Fortunately, the three-dimensional printer movement has created an entire community around designing the hardware and software to make this process easier. One important factor to be considered for the project is the size of the motor. In the stepper motor world, there is a standard, the **National Electrical Manufacturers Association (NEMA)** standard, which dictates the size and torque of stepper motors. For this project, you'll be using a **NEMA 17 size** stepper motor as it will supply the required torque.

There are two types of stepper motors, **unipolar** and **bipolar**. Unipolar stepper motors normally have a 4-pin connection to the motor, as shown in the following diagram:

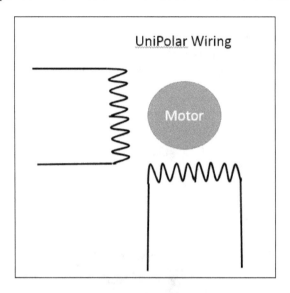

As you can see, unipolar motors don't have a center tap to the coil that is driving the magnetic fields, so they are a bit more difficult to control as the controller has to reverse the flow of current to step the motor. However, they are more powerful than bipolar motors.

Bipolar motors have a center tap to the coil that is driving the magnetic fields, as shown in the following diagram:

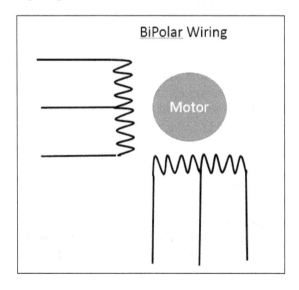

They are easier to control but have less torque. Fortunately, you can use bipolar motors as unipolar motors by simply connecting to the outside connections and ignoring the center tap. Since the controllers that you are going to use will take care of the driver's complexity, this is how you will control your stepper motors. Here is an image of the control pins on the stepper motor:

There are six pins. You will only use four pins to control the motor. If you don't have a wiring diagram for your motor, you'll need to discover which four wires you want to use. Use an **ohmmeter** to look for the two wires that are connected, that is, the two wires that don't show an infinite impedance. There should be three pins that are connected; however, one of the pins will show half of the impedance when connected to the other two wires in the set. This is the center tap wire, which you won't connect to the controller. In the end, you need two sets of wires that are connected to the outer wires of each of the coils.

In this case, for this particular stepper motor, the following are the proper connections:

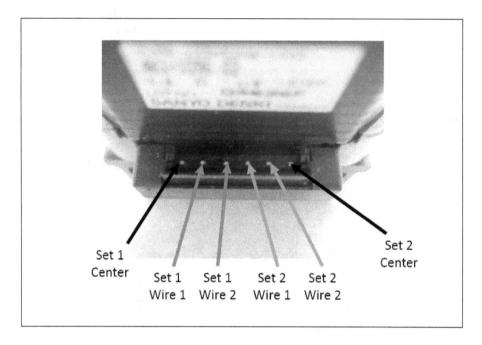

Now let's look at how to connect these to a stepper motor driver and Arduino. You are going to control the stepper motor by controlling the signals that you will send to these two wires. However, these signals are going to be quite large, so you'll need a stepper motor driver. You'll also need an interface board that allows you to connect the stepper motor driver to Arduino. Since you are using a three-dimensional printing setup, you'll use a complete system defined by that community. It is called **RepRap Arduino Mega Pololu Shield (RAMPS)**. For detailed information on this system, including its many suppliers, refer to http://www.reprap.org/wiki/RAMPS.

This system uses **Arduino Mega**, RAMPS, and up to five Pololu stepper motor drivers. Here is an image of the entire system:

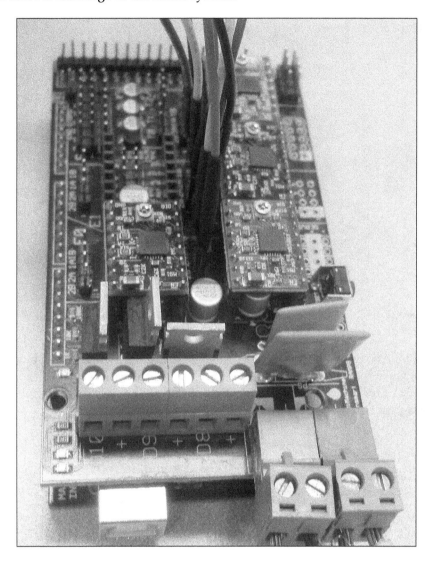

The lower board in the system is a standard Arduino Mega, one of the most powerful of the Arduino family. RAMPS is placed on top of Arduino and fits into the header connections. On top of RAMPS, there are up to five Pololu stepper motor drivers. Here is an image of an individual unit:

Each of these drives a separate stepper motor. In this case, you'll be driving three motors, so you'll need three of these drivers.

You'll also need a power supply that can supply the kind of voltage and current that you'll need to drive your stepper motors. Refer to `http://reprap.org/wiki/Power_supply` for the various options. One common choice is an OEM power supply that is designed to drive the LED light strips, similar to this one:

You have the basic system for driving stepper motors. Now you'll need to connect the RAMPS system to the motors. The connections for each motor are next to the stepper motor driver chip, as shown in the following image:

As you can see, there are four wires, two for each motor. You'll connect the first two wires to one winding and the second two to the other winding, similar to the following image:

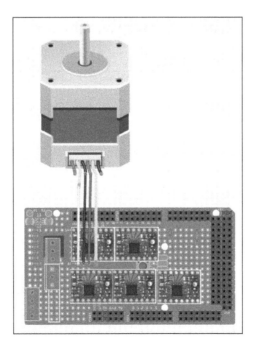

The first stepper motor driver is for the x-axis or moving the paddle back and forth across the table. You'll also need to connect the other two motors to the second and third set of stepper driver. The second set of wires is attached to the left motor (looking from behind the robot player) and the third set of wires is attached to the right motor.

Now that the connections are made, you'll need some software on the Arduino to send control signals to stepper motors. Fortunately, there is a code to test these connections and execute some basic movement of the paddle.

Moving the paddle with Arduino code

The first step in making the entire system work is to test the motors. Fortunately, the GitHub site, `https://github.com/JJulio/AHRobot`, has the code that can make this happen. Download and unzip the code. Look in the `AHRobot-master/Arduino/Utils/AHR_Motor_Test` directory for the `AHR_Motor_Test.ino` program file. This file provides a simple test program to move the three stepper motors, first the x-axis stepper motor that moves the paddle back and forth across the table and then the y-axis stepper motors, the two motors which move the paddle forward and backward.

 If you are unfamiliar with how to develop and upload code for Arduino Mega, go to the `https://www.arduino.cc/` website. It has a detailed set of instructions and an open source IDE for Arduino.

Run the program and open the **Serial Port**. You will see the following as the paddle moves:

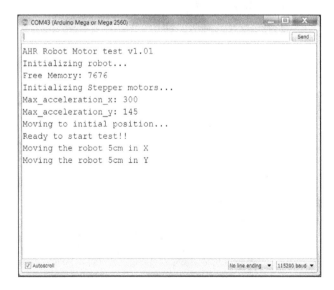

If everything is connected correctly, the paddle should first move across the table, then move forward. If it does not work, check the connections carefully.

The `Configuration.h` file is the only file that you'll need to change initially, it has some configuration values which you may need to adjust for your specific configuration. You may need to change the polarity of the motors to make sure that they run in the correct direction, this code is shown in the following figure:

```
AHR_Motor_Test | Arduino 1.6.1
File Edit Sketch Tools Help

AHR_Motor_Test   Configuration.h   Controller.h   Weights

//#define MAX_ACCEL_Y 100      //140//120
//#define MAX_SPEED_X 28000       //max 28000 for 12V   // Maximum speed in steps/seg
//#define MAX_SPEED_Y 28000


// This is for the Accel ramp implementation (to smooth the intial acceleration), simplified S-profile
#define ACCEL_RAMP_MIN 2500  // The S profile is generated up to this speed
#define ACCEL_RAMP_MAX 10000

// UNCOMMENT THIS LINES TO INVERT MOTORS
#define INVERT_X_AXIS 1
#define INVERT_Y_AXIS 1   //Y-LEFT
//#define INVERT_Z_AXIS 1   //Y_RIGHT
```

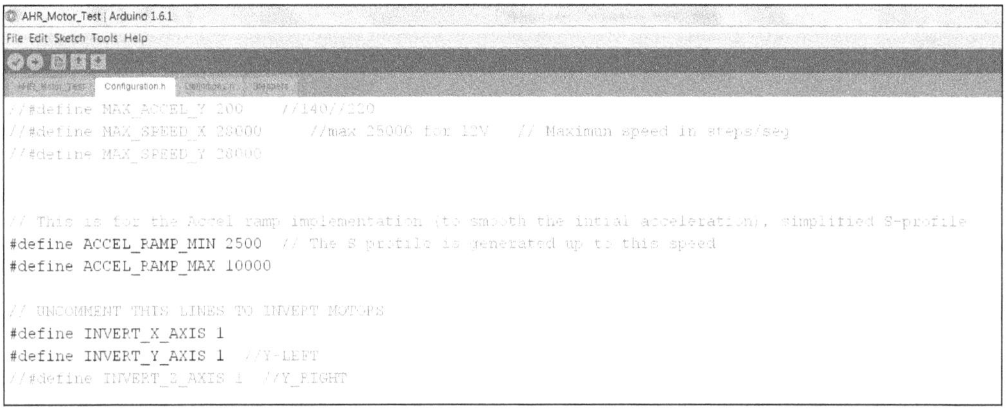

The other value that you may need to change is related to how far the paddle moves. In this case, the movement, as noted by the code, is dependent on the size of the gears. You may need to change this value, smaller values mean that the paddle will move shorter distances. Adjust this so that the paddle moves approximately two inches during the test, as shown in the following figure:

```
#define ACCEL_RAMP_MIN 2500  // The S profile is generated up to this speed
#define ACCEL_RAMP_MAX 10000


// UNCOMMENT THIS LINES TO INVERT MOTORS
#define INVERT_X_AXIS 1
#define INVERT_Y_AXIS 1   //Y-LEFT
//#define INVERT_Z_AXIS 1   //Y_RIGHT

// Geometric calibration.
// This depends on the pulley teeth. For 42 teeth GT2 => 18, for 40 teeth GT2 => 20, for 16 teeth T5 => 10
#define X_AXIS_STEPS_PER_UNIT 10     // With 42 teeth GT2 pulley and 1/8 microstepping on drivers
#define Y_AXIS_STEPS_PER_UNIT 10     // 200*8 = 1600 steps/rev = 1600/42teeth*2mm = 19.047, using 15 is an e

// Absolute Min and Max robot positions in mm (measured from center of robot pusher)
#define ROBOT_MIN_X 100
#define ROBOT_MIN_Y 80
#define ROBOT_MAX_X 500
#define ROBOT_MAX_Y 400
```

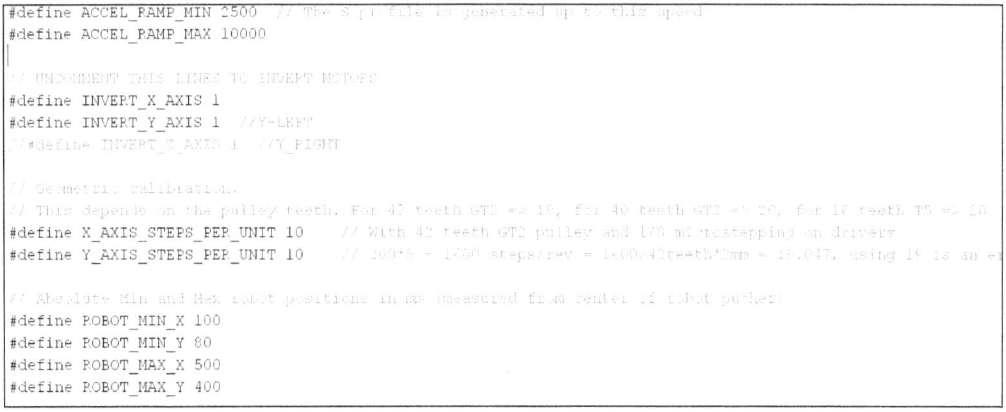

Finally, you'll want to edit the `Configuration.h` file to give it your specific table size. Here are these values:

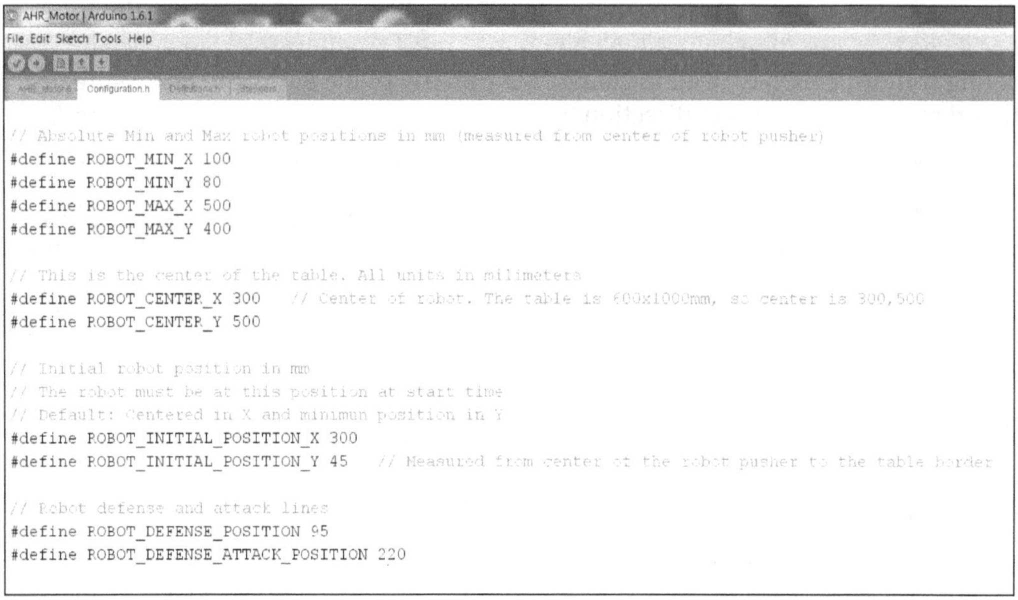

Now that you can move the paddle, you'll need to edit this code so that you can direct the paddle to a specific location. Let's start by changing the `loop()` function to move a short distance in *x* and *y* direction based on an input character. The code will look as shown:

```
void loop()
{
  int dt;
  uint8_t logOutput=0;
  debug_counter++;
  timer_value = micros();
  if ((timer_value-timer_old)>=1000)   // 1Khz loop
  {
    while (Serial.available()) {
      // get the new byte:
      char inChar = (char)Serial.read();
      switch (inChar){
        case 'a':
```

```
            Serial.println("Moving the robot 1cm in X");
            print_values();
            // We move the robot +1cm in X
            com_pos_x -= 10;
            setPosition(com_pos_x,com_pos_y);
            print_values();
            break;
            case 'd':
            Serial.println("Moving the robot -1cm in X");
            print_values();
            // We move the robot -1cm in X
            com_pos_x += 10;
            setPosition(com_pos_x,com_pos_y);
            print_values();
            break;
            case 'w':
            Serial.println("Moving the robot -1cm in Y");
            print_values();
            // We move the robot -1cm in Y
            com_pos_y += 10;
            setPosition(com_pos_x,com_pos_y);
            print_values();
            break;
            case 's':
            Serial.println("Moving the robot -1cm in Y");
            print_values();
            // We move the robot -1cm in Y
            com_pos_y -= 10;
            setPosition(com_pos_x,com_pos_y);
            print_values();
            break;
        }
    } // Serial input character loop
  positionControl();
  } // 1Khz loop
}
```

The operation is quite simple. If you send Arduino an *a* character through the **Serial Port**, it will move to the right; if you send it a *d* character, it will move to the left. A *w* character moves the unit forward, an *s* character moves it back.

What you are really going to want is a program where you will send it the *x* and *y* location and the paddle will move to that location. Here is the code for this action:

```
void loop()
{
  int dt;
  uint8_t logOutput=0;
  debug_counter++;
  timer_value = micros();
  if ((timer_value-timer_old)>=1000)   // 1Khz loop
  {
    char data[7];
    char x[4];
    char y[4];
    while (Serial.available()) {
      while (Serial.available() >= 7)
      {
        for(int i = 0 ; i < 7; i++)
          data[i] = Serial.read();//etc.
      }
      for(int i = 0 ; i < 3; i++)
        x[i] = data[i];
      x[3] = 0;
      for(int i = 4 ; i < 7; i++)
        y[i - 4] = data[i];
      y[3] = 0;
      com_pos_x = atoi(x);
      com_pos_y = atoi(y);
      print_values();
      setPosition(com_pos_x,com_pos_y);
      print_values();
    } // Serial input character loop
  positionControl();
  } // 1Khz loop
}
```

Now, you'll need to send the *x* and *y* values to the serial port to control your paddle using Raspberry Pi to determine where the puck is traveling. Now you will turn to Raspberry Pi to track the puck.

Seeing the puck using OpenCV

To know where the puck is, you'll need vision. Fortunately, adding hardware and software for vision for Raspberry Pi is both easy and inexpensive. First, you'll need to connect to a USB webcam.

Installing a USB camera on Raspberry Pi

Connecting a USB camera is very easy. Just plug it in the USB slot. To make sure your device is connected, type `lsusb`. You will see the following:

```
pi@raspberrypi ~ $ lsusb
Bus 001 Device 002: ID 0424:9514 Standard Microsystems Corp.
Bus 001 Device 001: ID 1d6b:0002 Linux Foundation 2.0 root hub
Bus 001 Device 003: ID 0424:ec00 Standard Microsystems Corp.
Bus 001 Device 004: ID 1a86:7523 QinHeng Electronics HL-340 USB-Serial adapter
Bus 001 Device 005: ID 413c:3012 Dell Computer Corp. Optical Wheel Mouse
Bus 001 Device 008: ID 046d:0825 Logitech, Inc. Webcam C270
Bus 001 Device 007: ID 413c:2003 Dell Computer Corp. Keyboard
pi@raspberrypi ~ $
```

This shows a **Logitech** webcam located at `Bus 001 Device 008: ID 046d:0825`. To make sure that the system sees this as a video device, type `ls /dev/v*` and you will see something similar to the following:

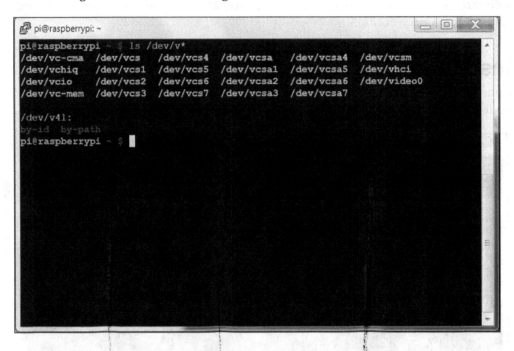

The `/dev/video0` is the webcam device. Now that your device is connected, let's see whether you can actually capture images and videos. There are several tools that can allow you to access the webcam, but a simple program with video controls is called **guvcview**. To install this, type `sudo apt-get install guvcview`. Once the application is installed, you'll want to run it. To do this, you'll either need to be directly connected to a display or access Raspberry Pi via a remote **VNC** connection, such as **VNC Server**, as displaying the images will require a graphical interface.

Once you are connected in this manner, open a terminal window on Raspberry Pi and run `guvcview -r 2`. You should see something similar to this:

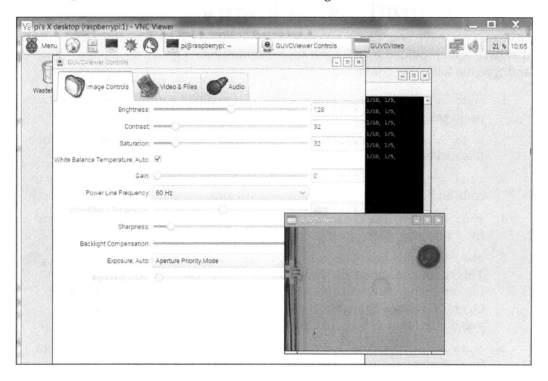

Don't worry about the resolution or quality of the image as you'll be capturing and processing your images inside OpenCV, a vision framework. You may also need to adjust the distance and the orientation of the webcam from the table.

Downloading and installing OpenCV – a fully featured vision library

Now that you have your camera connected, you can access the amazing capabilities that have been provided by the open source community. Open a terminal window and type the following commands:

1. `sudo apt-get update`: You're going to download a number of new software packages, so it is better to make sure that everything is up to date.

2. `sudo apt-get install build-essential`: Although, you may have done this earlier, the library is essential for building OpenCV.

3. `sudo apt-get install libavformat-dev`: This library provides a way to code and decode the audio and video streams.

4. `sudo apt-get install ffmpeg`: This library provides a way to transcode the audio and video streams.

5. `sudo apt-get install libcv2.4 libcvaux2.4 libhighgui2.4`: This command shows the basic OpenCV libraries. Note the number in the command. This will almost certainly change as the newer versions of OpenCV become available. If `2.4` does not work, then either try `3.0` or google the latest version of OpenCV.

6. `sudo apt-get install python-opencv`: This is the Python development kit that is needed for OpenCV, as you are going to use Python.

7. `sudo apt-get install opencv-doc`: This command will show the documentation for OpenCV, just in case you need it.

8. `sudo apt-get install libcv-dev`: This command shows the header file and static libraries to compile OpenCV.

9. `sudo apt-get install libcvaux-dev`: This command shows more development tools for compiling OpenCV.

10. `sudo apt-get install libhighgui-dev`: This is another package that provides the header files and static libraries to compile OpenCV.

11. `cp -r /usr/share/doc/opencv-doc/examples /home/pi/`: This command will copy all the examples to your home directory.

Now that OpenCV is installed, you can try one of the examples. Go to the `/home/pi/examples/python` directory. If you do an `ls`, you'll see a file named `camera.py`. This file has the most basic code to capture and display a stream of images. Before you run the code, make a copy of it, using `cp camera.py myCamera.py`. Then, edit the file to look as shown in the following:

```
pi@raspberrypi: ~/examples/python
File Edit Options Buffers Tools Python Help
import cv2.cv as cv
import time

cv.NamedWindow("camera", 1)

capture = cv.CaptureFromCAM(0)
cv.SetCaptureProperty(capture, 3, 360)
cv.SetCaptureProperty(capture, 4, 240)

while True:
    img = cv.QueryFrame(capture)
    cv.ShowImage("camera", img)
    if cv.WaitKey(10) == 27:
        break
cv.DestroyAllWindows()




-UU-:----F1  myCamera.py     All L6      (Python)----------------------------------
Wrote /home/pi/examples/python/myCamera.py
```

The two lines that you'll add are the two with the `cv.SetCaptureProperty()`
function, they will set the resolution of the image to 360 x 240. To run this program,
you'll need to either have a display and keyboard connected to Raspberry Pi or use
VNC Viewer. When you run the code, you will see the window displayed, as shown
in the following image:

This is the resolution that you'll use for this application. Just a note on the resolution. Bigger images are great—they give you a more detailed view of the world—but they take up significantly more processing power.

 For this application, you'll not want to use vncserver to display the images, this will also slow the system performance significantly. If you want to see the images in real time, connect a display, keyboard, and mouse to Raspberry Pi.

Color finding with OpenCV

Now you'll want to use OpenCV and your webcam to track your puck. OpenCV makes this amazingly simple by providing some high level libraries that can help you. To start with, you'll want to create a basic file that allows you to establish a threshold and then display the pixels as white that exceeds this threshold. To accomplish this, you'll edit a file to look something similar to what is shown in the following screenshot:

```
pi@raspberrypi: ~/examples/python
File Edit Options Buffers Tools Python Help
import numpy as np
import cv2

cap = cv2.VideoCapture(0)
cap.set(3,320)
cap.set(4,240)
low_range = np.array([10, 120, 100])
high_range = np.array([70, 255, 255])

while(cap.isOpened()):
    ret, frame = cap.read()
    hue_image = cv2.cvtColor(frame, cv2.COLOR_BGR2HSV)
    threshold_img = cv2.inRange(hue_image, low_range, high_range)
    cv2.imshow('video',frame)
    cv2.imshow('frame',threshold_img)
    if cv2.waitKey(1) & 0xFF == ord('q'):
        break

cap.release()
cv2.destroyAllWindows()

-UU-:----F1  try1.py        All L18    (Python)-----------------------------
Wrote /home/pi/examples/python/try1.py
```

Let's look specifically at the code that makes it possible to isolate the colored puck:

- `hue_img = cv.cvtColor(frame, cv.COLOR_BGR2HSV)`: This line creates a new image and stores it as per the values of **Hue** (color), **Saturation, and Value (HSV)** instead of the **Red, Green, and Blue (RGB)** pixel values of the original image. Converting to this format (HSV) focuses our processing more on the color as opposed to the amount of light hitting it.

- `threshold_img = cv.inRange(hue_img, low_range, high_range)`: The `low_range` and `high_range` parameters determine the color range. In this case, it is an orange ball so you want to detect the color orange. For a good tutorial on using hue to specify color, try `http://www.tomjewett.com/colors/hsb.html`. Also, `http://www.shervinemami.info/colorConversion.html` includes a program that you can use to determine your values by selecting a specific color.

Now, run the program. If you see a single black image window, move it and you will expose the original image window. Now take your target (in this case, the puck) and move it into the frame. You will see the following screenshot:

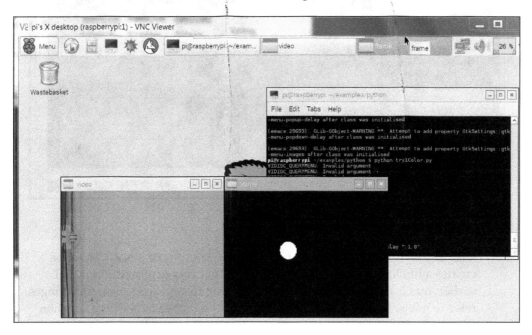

Notice the white pixels in our threshold image showing where the puck is located. You can add more OpenCV code that gives the actual x and y location of the puck. In the original image file of the puck's location, you can actually draw a circle around the puck as an indicator. Edit the file to look as shown in the following:

```
pi@raspberrypi: ~/examples/python
File Edit Options Buffers Tools Python Help
import numpy as np
import cv2

cap = cv2.VideoCapture(0)
cap.set(3,320)
cap.set(4,240)
low_range = np.array([10, 120, 100])
high_range = np.array([70, 255, 255])

while(cap.isOpened()):
    ret, frame = cap.read()
    hue_image = cv2.cvtColor(frame, cv2.COLOR_BGR2HSV)
    threshold_img = cv2.inRange(hue_image, low_range, high_range)
    contour, hierarchy = cv2.findContours(threshold_img, cv2.RETR_TREE, cv2.CHA\
IN_APPROX_SIMPLE)
    center = contour[0]
    moment = cv2.moments(center)
    (x,y),radius = cv2.minEnclosingCircle(center)
    center = (int(x),int(y))
    radius = int(radius)
    img = cv2.circle(frame,center,radius,(0,255,0),2)
    cv2.imshow('video',frame)
    if cv2.waitKey(1) & 0xFF == ord('q'):
        break


cap.release()
cv2.destroyAllWindows()

-UU-:**--F1  try1.py       All L22    (Python)------------------------------
```

The added lines look as follows:

- `hue_image = cv2.cvtColor(frame, cv2.COLOR_BGR2HSV)`: This line creates a hue image out of the RGB image that was captured. As noted earlier, hue is easier to deal with when trying to capture real-world images, refer to `http://www.bogotobogo.com/python/OpenCV_Python/python_opencv3_Changing_ColorSpaces_RGB_HSV_HLS.php` for details.

- `threshold_img = cv2.inRange(hue_image, low_range, high_range)`: This creates a new image that contains only those pixels that occur between the `low_range` and `high_range` n-tuples.

- `contour, hierarchy = cv2.findContours(threshold_img, cv2.RETR_TREE, cv2.CHAIN_APPROX_SIMPLE)`: This finds the contours or groups of like pixels in the `threshold_img` image.

- `center = contour[0]`: This identifies the first contour.

- `moment = cv2.moments(center)`: This finds the moment of this group of pixels.

- `(x,y),radius = cv2.minEnclosingCircle(center)`: This gives the x and y locations and `radius` of the minimum circle that will enclose this group of pixels.

- `center = (int(x),int(y))`: This finds the center of the x and y locations.

- `radius = int(radius)`: This converts radius of the circle to the integer type.

- `img = cv2.circle(frame,center,radius,(0,255,0),2)`: This draws a circle on the image.

Now that the code is ready, you can run it. You will see something similar to the image shown in the following screenshot:

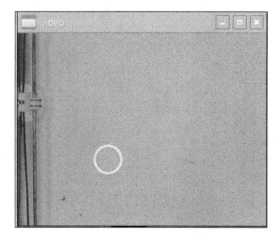

You can now find out the location of the puck. You also have the location of your puck, you'll use this to determine the location and direction of the puck.

Tracking the puck

Once you know the location, you can now find out the speed and direction of the puck. In this case, the easiest way is to find the delta movement in the x direction and the delta movement in the y direction. This is easy to add to your program by simply tracking your puck from frame to frame, that is, how much the puck has moved in pixels. Here is the code:

```python
import cv2
import numpy as np

cap = cv2.VideoCapture(0)
cap.set(3, 360)
cap.set(4, 240)
low_range = np.array([10, 120, 100])
high_range = np.array([70, 255, 255])
lastX = 0
lastY = 0
deltaX = 0
deltaY = 0

while (cap.isOpened()):
    ret, frame = cap.read()
    hue_image = cv2.cvtColor(frame, cv2.COLOR_BGR2HSV)
    threshold_img = cv2.inRange(hue_image, low_range, high_range)
    contour, hierarchy = cv2.findContours(threshold_img, cv2.RETR_TREE, cv2.CHAIN_APPROX_SIMPLE)
    if contour:
        center = contour[0]
        moment = cv2.moments(center)
        (x,y),radius = cv2.minEnclosingCircle(center)
        center = (int(x), int(y))
        deltaX = int(x) - lastX
        deltaY = int(y) - lastY
        lastX = int(x)
        lastY = int(y)
        radius = int(radius)
        img = cv2.circle(frame, center, radius, (0, 255, 0), 2)
        img = cv2.line(frame, (lastX,lastY), (lastX + deltaX, lastY + deltaY), (0, 255, 0), 2)
    cv2.imshow('video', frame)
    if cv2.waitKey(10) == 27:
        break

cap.release
cv2.destroyAllWindows()
```

-UU-:**--F1   **trackPuck.py**   All L32   (Python)--------------------------------

And when you run the code, you will see the following:

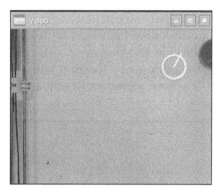

Here, the line is pointing in the direction of the puck movement. Now that you know the location and direction of the puck, you can move the paddle to connect with the puck.

Moving the paddle to strike the puck

You'll need to add some commands to the puck tracking program to talk over the serial port and move the paddle. But first, let's explore how to send some simple commands from inside a python program that is running on Raspberry Pi to control the paddle. Here is a simple program that takes in user input and sends it to the Arduino control program:

```
pi@raspberrypi: ~/examples/python
File Edit Options Buffers Tools Python Help
#!/usr/bin/python
import serial
import time

ser = serial.Serial('/dev/ttyACM0', 115200, timeout = 1)

while 1:
    x = raw_input("Enter x and y value, xxx,xxx format: ")
    print x
    ser.write(x)
    time.sleep(1)




-UU-:----F1  simpleSerial.py   All L9     (Python)-------------------
Wrote /home/pi/examples/python/simpleSerial.py
```

When you run this code, you will be able to enter a location and the paddle will go to that location. Now you'll want to tie this code in to the code for puck tracking; however, you'll need a function that can calculate where the paddle needs to be, based on the puck speed and direction. Let's start with a very easy function, one that assumes no bounce on the side and that the paddle will stay at y = 0. As an example, let's look at when the puck is at a location with an equal x and y velocity. The following is a diagram of where you want to move the puck:

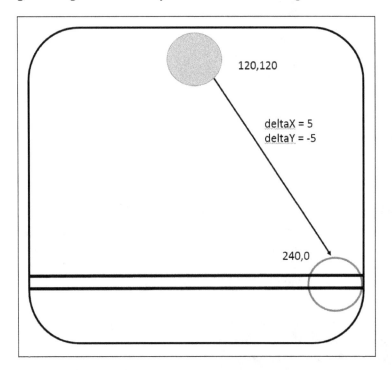

You'll also need to add the idea of bounce off the edge of the table. Adding the bounce is actually quite easy, if the value of x is less than zero or greater than the maximum value of the table, then you'll either add to zero (if less than zero) or subtract from the maximum value of the table as the bounce will come off at the same angle as it goes in. The code will look similar to the following:

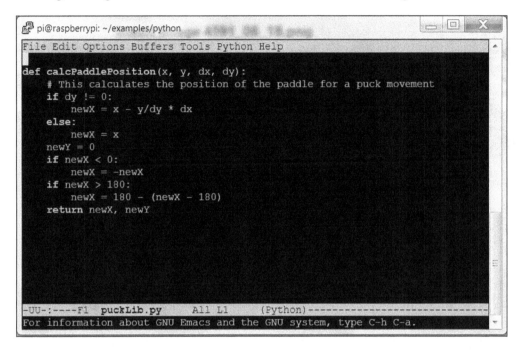

```
def calcPaddlePosition(x, y, dx, dy):
    # This calculates the position of the paddle for a puck movement
    if dy != 0:
        newX = x - y/dy * dx
    else:
        newX = x
    newY = 0
    if newX < 0:
        newX = -newX
    if newX > 180:
        newX = 180 - (newX - 180)
    return newX, newY
```

The 180 pixels, in this case, is the maximum x value for this table. You'll want to change this if you are using a higher resolution and a bigger table. Now you can import this function as a library into your puck tracking code by removing everything except the function code and you will get the proper puck movement. This code will look similar to the following screenshot:

```
pi@raspberrypi: ~/examples/python
File Edit Options Buffers Tools Python Help
import cv2
import numpy as np
from puckLib import *
import time
import serial
ser = serial.Serial('/dev/ttyACM0', 115200, timeout = 1)
cap = cv2.VideoCapture(0)
cap.set(3, 180)
cap.set(4, 120)
low_range = np.array([10, 120, 100])
high_range = np.array([70, 255, 255])
lastX = 0
lastY = 0
deltaX = 0
deltaY = 0
while (cap.isOpened()):
    ret, frame = cap.read()
    hue_image = cv2.cvtColor(frame, cv2.COLOR_BGR2HSV)
    threshold_img = cv2.inRange(hue_image, low_range, high_range)
    contour, hierarchy = cv2.findContours(threshold_img, cv2.RETR_TREE, cv2.CHAIN_APPROX_SIMPLE)
    if contour:
        center = contour[0]
        moment = cv2.moments(center)
        (x,y),radius = cv2.minEnclosingCircle(center)
        y = 120 - y
        center = (int(x), int(y))
        deltaX = int(x) - lastX
        deltaY = int(y) - lastY
        lastX = int(x)
        lastY = int(y)
        radius = int(radius)
        img = cv2.circle(frame, (lastX, 120 - lastY), radius, (0, 255, 0), 2)
    cv2.imshow('video', frame)
    if deltaY < 1:
        newPaddleX, newPaddleY = calcPaddlePosition(lastX, lastY, deltaX, deltaY)
        xCommand = str(newPaddleX)
        yCommand = str(newPaddleY)
        command = xCommand.zfill(3) + ',' + yCommand.zfill(3)
        ser.write(command)
        time.sleep(.2)
    if cv2.waitKey(10) == 27:
        break
cap.release
cv2.destroyAllWindows()
-UU-:**--F1   trackPuck.py   All L9       (Python)-------------------------------
```

That's it, you will now be able to play with your robot. At this point, it will only move in the x direction, you can add the capability of moving in the y direction to add more power to its response by timing the puck and moving the y-axis as the puck arrives.

Summary

You now know how to control stepper motors in order to control an air hockey paddle. In the final chapter, you'll learn how to integrate Raspberry Pi into a quadcopter platform, making the sky your last robotic conquest.

7
A Robot That Can Fly

You've had the opportunity to build lots of different types of robots, so now let's end with one that can be truly amazing, a robot that can fly.

In this chapter, you'll learn the following:

- Building the basic quadcopter platform
- Interfacing Raspberry Pi to the flight controller
- Discussing long range communications
- Using GPS for location
- Adding autonomous flight

Constructing the platform

Constructing the quadcopter hardware can be daunting; however, there are several excellent websites that can lead you through the process from component selection to build details and programming and controlling your quadcopter with a radio. The `http://www.arducopter.co.uk/` website is a great place to start for those who are new to quadcopter flight. Go to `http://copter.ardupilot.com/`, which is another excellent website with lots of information.

For this project, you'll want to choose a project that uses the **Pixhawk** flight controller. There are other flight controllers that are significantly less expensive, but this particular flight controller provides easy access for Raspberry Pi. Here are some possible websites that can guide you through the construction process; `http://learnrobotix.com/uavs/quadcopter-build/pixhawk/connecting-the-q-brain-esc.html`, `http://www.thedroneinfo.com/2015/06/06/build-a-quadcopter-with-pixhawk-flight-controller/`, and `http://www.flying-drone.co.uk/how-to-build-a-quadcopter-with-a-pixhawk-flight-controller-step-11/`.

At `http://copter.ardupilot.com/wiki/advanced-pixhawk-quadcopter-wiring-chart/`, you'll find an excellent wiring diagram of how to hook everything up. Let's go through the steps of constructing our own quadcopter.

First, you'll need a frame. You'll be building a quadcopter of size 450 mm, one of the least expensive frames, which are available at most online retailers, with fiberglass arms, as shown in the following image:

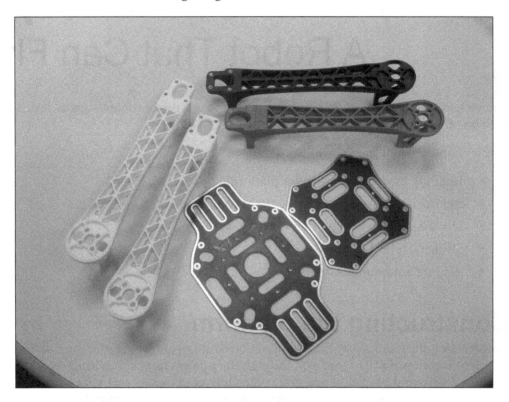

Now, follow the steps to complete your quadcopter assembly:

1. The first step is to build the quadcopter as the instructions suggest.

2. The second step is to solder the four **Electronic Speed Controllers (ESC)**, one to each motor, and the battery connection to the bottom plate. Here is an image of the bottom plate:

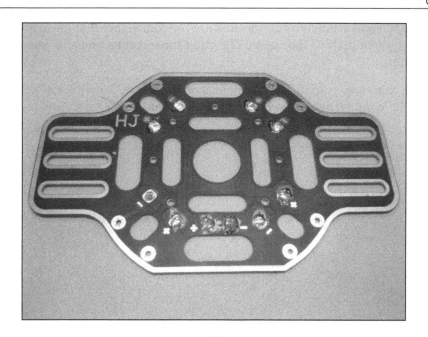

Notice the + and – connections; each connection will be soldered to all the ESCs. The following is an image of the motor controller:

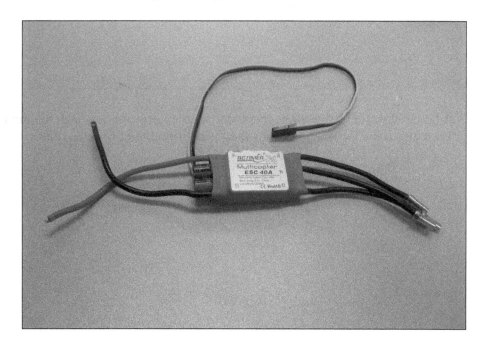

The red and white wire connectors are the connectors that are soldered to the bottom plate of the frame. The other three connectors will connect to the motor.

3. The third step is to install the motors on the frame. You'll want motors in the 1000KV range, here is an image of such a motor:

Again, follow the instructions that came with your frame to attach the motor. Then attach the three connections that come from the ESC to the motor.

4. One optional step is to add a landing gear set to the unit. There are many of these available. Here is an image of one that is very sturdy:

5. Now you'll install Pixhawk on the frame and connect its associated electronics. The details are shown and described at `http://copter.ardupilot.com/wiki/advanced-pixhawk-quadcopter-wiring-chart/`. This will connect the Pixhawk to the ESCs, the battery, an RC transmitter, a telemetry radio, and a switch that will prevent the quadcopter from flying until you are ready.

6. Eventually, you will install four propellers on the quadcopter; however, you will have to wait until you have calibrated the ESCs, motors, and RC transmitter to install them. You'll need four propellers, two that are designed to spin clockwise and other two that are designed to spin counter-clockwise. For this quadcopter, you'll want propellers that are 10 x 4.7 pitch. Here is an image of one such propeller:

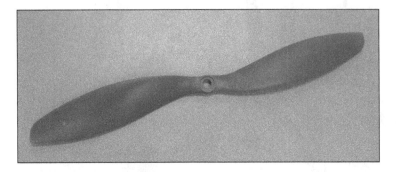

The following is an image of the entire quadcopter using the Pixhawk flight controller:

You'll notice the green arrows and cord arranged on the quadcopter. This is not to make it look menacing but to protect it from running into something and fracturing the propellers. There are commercial guards available; however, this system also works and is less expensive.

You'll want to build your quadcopter and fly it a bit with an RC transmitter/receiver pair. This will allow you to get familiar with your quadcopter and how it flies. It will also allow you to tweak all the settings to stabilize it. Once your quadcopter is stable, you can perform some simple autonomous flights. Let's start using the mission planning software, which runs on a remote computer.

Mission planning software

The mission planning software is available at `http://planner.ardupilot.com/`. There are actually two applications available that perform similar actions, but the **Mission Planner** is a good place to become familiar with how to talk with your quadcopter from a computer program.

To do this, you'll need to make sure you have telemetry radios connected to the Pixhawk and the computer. This will prevent the need of directly connecting to the Pixhawk with a long USB cable. When you begin the mission planning software, you will see the following screen:

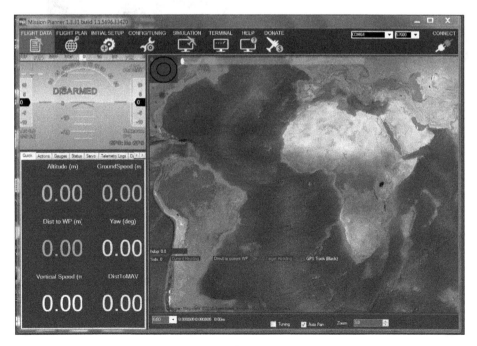

This is the basic screen. You'll then need to configure your radio's COM port and then press the **CONNECT** button in the corner on the upper right-hand side. As you move the quadcopter around, you will see the measurements change. If you are having problems connecting to the Pixhawk, there is lot of help available at the website.

Now that you have connected, you can actually see how your quadcopter is flying from this application. The software communicates with the Pixhawk controller via the **MAVLink**, a serial control link that comes from the software application, goes out over the telemetry radio, is received by the telemetry radio, and then is routed to the Pixhawk. The Pixhawk knows not only how to send information but also how to receive information.

Once the software is connected, you'll want to calibrate the RC radio connection. This can be done through the software. You'll also want to calibrate the ESCs, refer to http://learnrobotix.com/uavs/quadcopter-build/pixhawk/calibrating-electronic-speed-controllers-with-pixhawk.html for specific directions.

Now you are ready to connect Raspberry Pi. To do this, connect Raspberry Pi to the second telemetry input on the Pixhawk, as shown in the following:

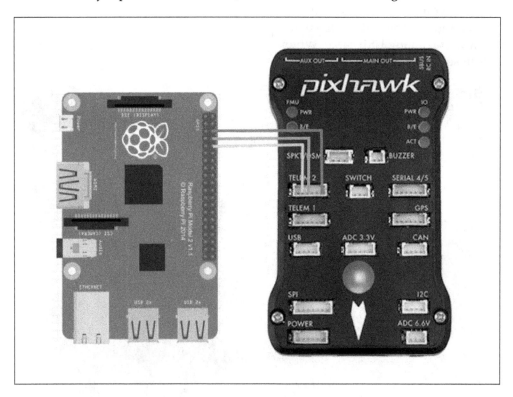

Now that this is connected, you can access the Pixhawk from Raspberry Pi using the MAVLink. Now, you'll need to add and configure the Raspberry Pi to complete the connection. To do this, run `raspi-config`, and choose the **8 Advanced Options**, **Configure advanced settings** selection, as shown in the following:

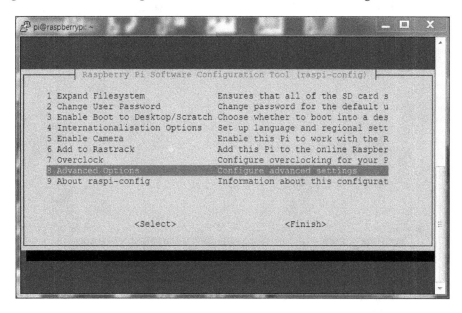

Now, you'll turn off sending the serial output on boot up by selecting the **A8 Serial**, **Enable/Disable shell and kernel m** option, as follows:

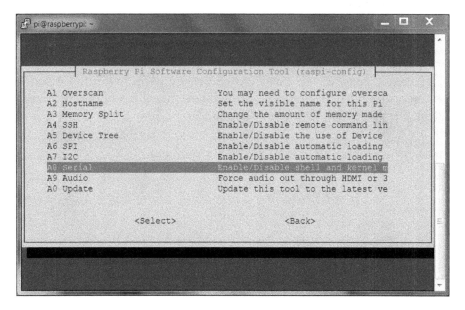

Then select the answer **<No>** to the following question:

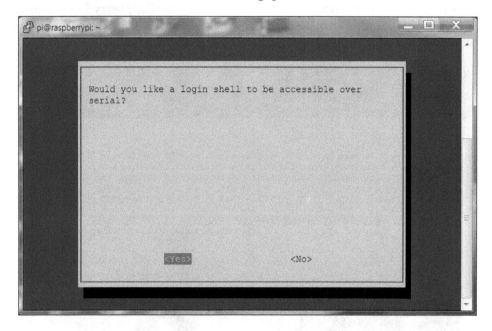

Now you are ready to install some additional software. To install this software, perform the following steps:

1. Type `sudo apt-get update`: This updates the links so that your system can find the appropriate software.

2. Type `sudo apt-get install screen python-wxgtk2.8 python-matplotlib python-opencv`: This installs a graphical package, a plotting package, and a version of OpenCV.

3. Type `sudo apt-get install python-numpy`: This will install **NumPy**, a numerical library for python, although you may already have it from the previous projects that you have done.

4. Type `sudo apt-get install python-dev`: This is a set of files that will allow you to develop in the Python environment.

5. Type `sudo apt-get install python-pip`: This is a tool that helps you install python packages.

6. Type `sudo pip install pymavlink`: This is the set of code that implements the MAVLink or the communication profile to the Pixhawk, in python.

7. Type `sudo pip install mavproxy`: This last step installs the **Unmanned Aerial Vehicle (UAV)** ground station software package for MAVLink based systems that are based on the Pixhawk.

Now that you have installed all the software, you can test the link. To do this, type `sudo -s`; this establishes you as the superuser. Then type `mavproxy.py --master=/dev/ttyAMA0 --baudrate 57600 --aircraft MyCopter` and you will see the following:

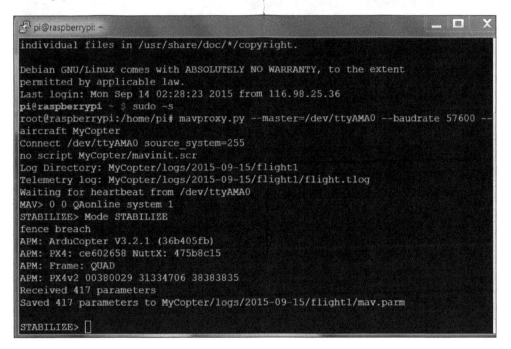

Now that the link is established, you can send commands to either set or show parameters. For example, type param show ARMING_CHECK, which should show you the value of the parameter, as shown in the following:

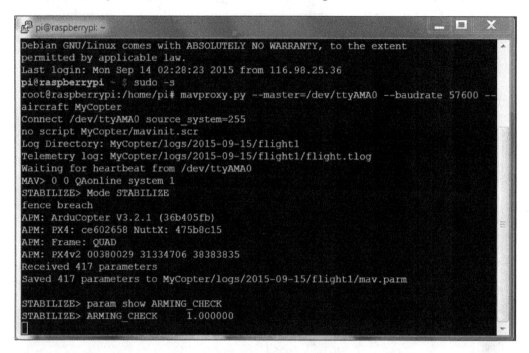

Details for all the commands available can be found at http://dronecode.github. io/MAVProxy/html/uav_configuration/index.html.

You can issue these commands directly, but you can also connect to the Pixhawk using an interface that is similar to the Mission Planner interface, which you worked with earlier. To do this, you'll need to install the **DroneKit** code. Overall directions and documentation for DroneKit can be found at http://python.dronekit.io/ guide/getting_started.html#installing-dronekit, but let's show an example here.

First, type sudo pip install droneapi. You can download some example scripts by typing git clone http://github.com/dronekit/dronekit-python.git. Now cd to the dronekit-python/examples/vehicle_state directory. You'll see the vehicle_state.py file that shows an excellent example of how to use the MAVLink to talk with the Pixhawk to find out information as well a set values and issue commands.

To run an example program, start the MAVLink by typing two commands: `sudo -s`, and then `mavproxy.py --master=/dev/ttyAMA0 --baudrate 57600 --aircraft MyCopter`. Once inside, load the API by typing `module load droneapi.module.api` at the prompt. The system will then tell you whether the module is loaded. Now, run the python script by typing `api start vehicle_state.py`.

The python code will first read in a series of parameters and then, if the quadcopter is armed, it will also read some details about the state of the quadcopter. Details of each command can be found at `http://python.dronekit.io/guide/vehicle_state_and_parameters.html#vehicle-information`. The output will look something like the following:

```
pi@raspberrypi: ~/dronekit-python/examples/vehicle_state

STABILIZE> module load droneapi.module.api
STABILIZE> DroneAPI loaded
Loaded module droneapi.module.api

STABILIZE> api start vehicle_state.py
STABILIZE>
Get all vehicle attribute values:
 Location: Location:lat=0.0,lon=0.0,alt=1.38999998569,is_relative=False
 Attitude: Attitude:pitch=0.0657835155725,yaw=-3.04151630402,roll=-0.02454243041
57
 Velocity: [0.0, 0.0, 0.0]
 GPS: GPSInfo:fix=0,num_sat=0
 Groundspeed: 0.0
 Airspeed: 0.0
 Mount status: [None, None, None]
 Battery: Battery:voltage=0.0,current=None,level=None
 Rangefinder: Rangefinder: distance=None, voltage=None
 Rangefinder distance: None
 Rangefinder voltage: None
 Mode: STABILIZE
 Armed: False
Set Vehicle.mode=GUIDED (currently: STABILIZE)
 Waiting for mode change ...
Got MAVLink msg: COMMAND_ACK {command : 11, result : 0}
APM: PreArm: Need 3D Fix
GUIDED> Mode GUIDED
Set Vehicle.armed=True (currently: False)
 Waiting for arming...
Got MAVLink msg: COMMAND_ACK {command : 400, result : 3}
 Waiting for arming...
```

Now, you can look at other python examples to see how to control your quadcopter via python files from Raspberry Pi.

You can also interface the MAVProxy system with the Mission Planner running on a remote computer. With a radio connected to the **TELEM 1** port of the Pixhawk and your Raspberry Pi connected to the **TELEM 2** port of the Pixhawk, change the MAVProxy startup command by adding `--out <ipaddress>:14550` with `ipaddress` being the address of the remote computer that is running the Mission Planner. On a Windows machine, the `ipconfig` command can be used to determine this IP address.

For example, your `mavproxy` command might look similar to this: `mavproxy.py --master=/dev/ttyAMA0 --baudrate 57600 --out ipaddress:14550 --aircraft MyCopter`. Once connected to MAVProxy, you can connect to the Mission Planner software using the UDP connection, as shown in the following screenshot:

Now, you can run your MAVProxy scripts and see the results on the Mission Planner software.

Summary

That's it. You now have a wide array of different robotics platforms that run with Raspberry Pi as the central controller. These chapters have just introduced you to some of the most fundamental capabilities of your platforms, you can now explore each and expand their capabilities. The only limits are your imagination and time.

Index

Raspberry Pi Robotic Projects

ISBN: 978-1-84969-432-2   Paperback: 278 pages

Create amazing robotic projects on a shoestring budget

1. Make your projects talk and understand speech with Raspberry Pi.

2. Use standard webcam to make your projects see and enhance vision capabilities.

3. Full of simple, easy-to-understand instructions to bring your Raspberry Pi online for developing robotics projects.

Raspberry Pi Robotics Essentials

ISBN: 978-1-78528-484-7   Paperback: 158 pages

Harness the power of Raspberry Pi with Six Degrees of Freedom (6DoF) to create an amazing walking robot

1. Construct a two-legged robot that can walk, turn, and dance.

2. Add vision and sensors to your robot so that it can "see" the environment and avoid barriers.

3. A fast-paced, practical guide with plenty of screenshots to develop a fully functional robot.

Raspberry Pi Blueprints

ISBN: 978-1-78439-290-1   Paperback: 284 pages

Design and build your own hardware projects that interact with the real world using the Raspberry Pi

1. Interact with a wide range of additional sensors and devices via Raspberry Pi.

2. Create exciting, low-cost products ranging from radios to home security and weather systems.

3. Full of simple, easy-to-understand instructions to create projects that even have professional-quality enclosures.

Learning Raspberry Pi

ISBN: 978-1-78398-282-0   Paperback: 258 pages

Unlock your creative programming potential by creating web technologies, image processing, electronics- and robotics-based projects using the Raspberry Pi

1. Learn how to create games, web, and desktop applications using the best features of the Raspberry Pi.

2. Discover the powerful development tools that allow you to cross-compile your software and build your own Linux distribution for maximum performance.

3. Step-by-step tutorials show you how to quickly develop real-world applications using the Raspberry Pi.

Please check **www.PacktPub.com** for information on our titles

www.ingramcontent.com/pod-product-compliance
Lightning Source LLC
Chambersburg PA
CBHW060129060326
40690CB00018B/3802